Sharing the Effort

During the financial year 1996/7 the Energy and Environmental Programme is supported by generous contributions of finance and technical advice from the following organizations:

Ashland Oil • British Gas • British Nuclear Fuels • British Petroleum
Brown and Root • Department of Trade and Industry • Eastern Electricity
Enron • Enterprise Oil • Esso/Exxon • LASMO • Magnox Electric
Mitsubishi Fuels • National Grid • Nuclear Electric • PowerGen • Ruhrgas
St Clements Services • Saudi Aramco • Shell • Statoil • TEPCO • Texaco
Veba Oil

Additional support is received for specific research projects from the following:

ABB • Chubu Electric Power Co • Commission of the European Communities
Department of Environment UK • East Asia Gas Company • ENEL
Hydro-Quebec • Imatran Voima Oy • Japan National Oil Company
London Electricity • Royal Ministry of Foreign Affairs, Norway
Ministry of Industry and Energy, Norway • Palmco • Siemens
Tennessee Valley Authority • UK Foreign and Commonwealth Office

Additional funding for specific meetings and workshops has been received from:

Amerada Hess • Department of Environment, Sport and Territories, Australia
Environment Agency, Japan • French Environment Ministry • GISPRI
Greenpeace • International Climate Change Partnership • Mobil
Rolls Royce Industrial Power • RTZ-CRA Group • Swiss Environment
Ministry • Uranium Institute US State Department

Sharing the Effort

Analysing Options for Differentiating
Commitments under the Framework
Convention on Climate Change

Report of a workshop held at
The Royal Institute of International Affairs
London, June 1996

edited by Matthew Paterson
and Michael Grubb

THE ROYAL INSTITUTE OF
INTERNATIONAL AFFAIRS
Energy and Environmental Programme

First published in Great Britain in 1996 by

Royal Institute of International Affairs, 10 St James's Square, London SW1Y 4LE

(Charity Registration No. 208 223)

Distributed in North America by

The Brookings Institution, 1775 Massachusetts Avenue NW,

Washington DC 20036-2188

A catalogue record for this book is available from the British Library.

Paperback: ISBN 1 89965 875 0

Printed and bound by The Chameleon Press Ltd.

Cover by Visible Edge.

Contents

List of figures and tables

Preface

In June 1996, the Royal Institute of International Affairs organized a workshop which was designed to explore the issue of whether future commitments under the climate convention should be differentiated between countries, and if so, how such differentiation could be developed. Participants in the workshop included policy-makers and representatives of many countries involved in international climate negotiations, as well as members of research institutes, non-governmental organizations and private industry. The aim was to inform and perhaps influence the course of debates on this subject leading up to the third Conference of the Parties in 1997. The timing of the workshop a month before the second Conference of the Parties was designed to help stimulate debate on this question.

Consistent with RIIA tradition and in the interest of encouraging uninhibited debate and a full airing of views, the workshop was conducted under the Chatham House Rule, meaning participants' remarks cannot be attributed. Consequently, the only names which appear here are those of the people who subsequently produced written papers for the workshop.

Following our practice for earlier workshops on climate change negotiations and policies,* this report has been produced from the workshop in order to inform a broader audience of the debates discussed. It contains selected papers presented at the workshop, and a Rapporteur's report which provides an overview of the discussions in which participants in the workshop engaged.

---

* Michael Grubb and Nicola Steen (eds), *Pledge and Review Process: Report of a Workshop*, RIIA, London, 1991. Pier Vellinga and Michael Grubb (eds), *Climate Change Policy in the European Community: Report of a Workshop*, RIIA, London, 1993. Michael Grubb and Dean Anderson (eds), *The Emerging International Regime for Climate Change: Structures and Options After Berlin: Report of a Workshop*, RIIA, London, 1995.

The Rapporteur's report is organized thematically, rather than strictly following the debates as they developed throughout the two days. This is to provide a balance between the need to give a coherent overview of the discussions and the need to ensure confidentiality for the workshop participants. The Rapporteur's report is followed by seven of the papers presented to the workshop. These were selected according to their inherent interest and the way in which they reflect the breadth of the themes raised at the workshop, although also of course we were dependent on the willingness of the authors to produce written versions of their papers in a short period of time. We hope that we have been able to do justice to the discussions, and that this report stimulates the debate on what may prove to be an extremely thorny issue in climate change negotiations.

November 1996   Matthew Paterson, Rapporteur
                                   Michael Grubb
              Head, Energy and Environmental Programme
              Royal Institute of International Affairs

Acknowledgments

We are immensely grateful to all the contributors to this report. We are also grateful to all those who attended the workshop and offered their insights on the many difficult issues raised by the question of differentiation, and for contributing to what proved a very interesting and wide-ranging set of discussions.

We are extremely grateful to the following for providing the financial support which made this workshop possible:

Department of Environment, Sport and Territories, Australia;
The Environment Agency, Japan;
Global Industrial and Social Progress Research Institute (GISPRI), Japan;
Ministry of Environment, France;
The Royal Ministry of Foreign Affairs, Norway;
The State Department, USA.

We are also grateful for the intellectual input of individuals in these and other departments, as well as advice from the Secretariat to the Climate Convention. The contents of this report remain, of course, the responsibility of the editors and named authors.

November 1996   Michael Grubb

Abbreviations

AGBM	Ad Hoc Group on the Berlin Mandate
CFCs	Chlorofluorocarbons
COP	Conference of the Parties
EIT	Economies in Transition
GDP	Gross domestic product
GEF	Global Environment Facility
GHGs	Greenhouse gases
GNP	Gross national product
HCFCs	Hydrochlorofluorocarbons
JUSCANZ	Japan, USA, Canada and New Zealand
LCP	Large combustion plant
LNG	Liquefied natural gas
LRTAP	Long Range Transboundary Air Pollution (agreement)
NGO	Non-governmental organization
NIC	Newly industrializing country
ODS	Ozone-depleting substances
OECD	Organization for Economic Cooperation and Development
OPEC	Oil Producing and Exporting Countries
PPP	'Polluter pays' principle
QELRO	Quantified Emission Limitation and Reduction Objective
REEO	Reduction of Excess Emissions Objective
UNDP	United Nations Development Programme
UNEP	United Nations Environment Programme
UNFCCC	United Nations Framework Convention on Climate Change
USTIES	US Technologies for International Environmental Solutions
VOC	Volatile Organic Compound

Chapter 1

Introduction and Rapporteur's report of workshop presentations and discussions

Matthew Paterson

Introduction

The UN Framework Convention on Climate Change (UNFCCC) recognizes that countries bear 'common but differentiated responsibilities' in addressing climate change, and divides commitments explicitly between 'Annex I' and 'non-Annex I' countries.[1] In the negotiations on climate change following its signing, and more particularly since the first Conference of the Parties to the Convention (COP) in Berlin in 1995, some suggested that future commitments should be differentiated further. The need to take countries' specific situations into account in future negotiations was reflected in paragraph 2(a) of the Berlin Mandate. During debates on this question, a number of positions have emerged.

The issue of differentiation of future commitments between and within groups of countries under the Climate Change Convention has therefore emerged, and it looks as if it could become one of the thorny issues within climate negotiations. Yet it is also one which has had little open discussion. To provide a forum for exploring this issue, the Royal Institute of International Affairs organized a workshop designed to explore possible ways in which commitments could be differentiated, should states choose to differentiate their commitments further in the future.[2] This report is based on the discussions at the workshop, and is followed by some of the papers presented there.

---

[1] See *United Nations Framework Convention on Climate Change*, United Nations, New York, 1992, Articles 3.1 (on the principle of common but differentiated responsibility) and 4 (on the differentiation of commitments between industrialized and developing countries).
[2] The workshop focused primarily on differentiation of commitments among Annex I countries.

To differentiate or not to differentiate

Some states have been vocal in asserting that any future commitments should be differentiated more than is currently the case, arguing that it would be desirable to have differentiation of commitments within Annex I countries, and perhaps within non-Annex I countries.

Several general arguments are presented for differentiation of commitments:

- *Economic:* differentiation of commitments could be considerably more efficient than flat-rate emissions reductions, since countries in which emissions reductions are cheaper could do more than others.
- *Ethical:* differentiation allows for a fairer distribution of the burden of reducing emissions. This could be applied regardless of whether one considers fairness in distribution burdens to be based on responsibility for causing climate change, or on ability to pay, or on any other criteria.
- *Pragmatic:* since states will in practice only sign up to agreements which contain commitments they are willing to meet, differentiation of those commitments could take account of this fact.
- *In terms of effectiveness:* differentiation could facilitate an agreement whereby all parties would be required to take action beyond business-as-usual emissions and 'no regrets'.

At the other end of the spectrum, some states adhere to the position that differentiation would only complicate negotiations further, and that simple, straightforward commitments such as uniform targets remain the best way forward at the present time, perhaps with Joint Implementation to allow for some flexibility in meeting commitments. Some at the workshop suggested that the Convention only refers to differentiation of responsibilities, not of commitments. Others argued that the emergence of differentiation as an issue could be a strategy to slow down negotiations under the Berlin Mandate, and could lead to special pleading to allow some states to get out of more stringent commitments on greenhouse gas emissions.

In between, a number of positions could be identified. One, expressed widely at the workshop, was that although differentiation of commitments was probable in the future, it was unlikely that states would be able to devise credible means of differentiating commitments before the third Conference of the Parties (COP 3) in December 1997, the timescale specified by the Berlin Mandate for renegotiating commitments. The best option for developing differentiated commitments, in this view, was to begin an extensive dialogue on possible bases for differentiation further into the future.

This became a widely shared – but not universal – view at the workshop. Most participants believed it was unlikely that any politically feasible scheme for differentiating commitments would emerge before COP 3, although some also believed this would mean that commitments negotiated by then would be weaker than they otherwise could be. A strong view emerged that in the longer term greater differentiation of commitments was likely – especially as states grow progressively more concerned with the distribution of the burden as commitments to reduce greenhouse gas emissions become more stringent. Therefore, the focus for present work on this theme should be on possible means of differentiating commitments which might be acceptable to countries beyond COP 3. The workshop discussed a broad range of these means of differentiating commitments in that longer timeframe.

Differentiation in the Convention

A clear account of how commitments are differentiated in the Framework Convention was given at the workshop in the paper by Kinley and Terrill (Chapter 2). On behalf of the UNFCCC secretariat, they suggested that there are two general kinds of differentiation which have emerged in discussions of the concept: concrete differentiation, where different countries or groups of countries have different commitments; and contextual differentiation, where countries have the same commitments, but they are written in such a way as to allow flexibility in their implementation, taking account of national circumstances.

The paper then went on to detail a substantial number of ways in which the Convention allows for differentiation of commitments, the clearest being between Annex I countries and others, and Annex II countries and others. This is a clear example of concrete differentiation. However, there is a significant number of examples of contextual differentiation in the Convention, such as those articles which allow some flexibility to some parties in meeting their commitments, or which make allowances for the special circumstances of a large variety of types of countries.

The Convention in general also explicitly makes allowance for differentiation of commitments, as mentioned above. However, there was a debate in the workshop about how this should be interpreted. It was argued that the possibility of differentiation should not be taken to mean allowing some countries to evade their responsibility in dealing with climate change. Others had a different interpretation of 'common but differentiated responsibility', arguing that responsibility in terms of contributing to climate change did not necessarily translate directly into responsibility to act to mitigate climate change in the future, and that the Convention explicitly mentions differentiation of responsibilities, but defines differentiation of commitments less clearly.

Approaches to differentiation

The question of differentiation of commitments is fundamentally about equity. However, as has been widely noted, there are many different conceptions of what equity means in practice.[3] In the context of this discussion, this produces a number of positions.

One option which, as noted above, is likely to dominate unless governments can develop credible alternatives, is equal reductions. While it may have an intuitive ring of fairness, it is not widely seen to be fair by those who have examined it closely. This is one of the motives behind discussing alternatives. However, it has much of the force of precedent behind it, and is feasible both to negotiate and to implement. Thus it remains the default position in the negotiations, the preferred option for many states and for some participants in the workshop. Further differentiation will only happen if credible alternative proposals are put on the negotiating table.

---

[3] For discussions of the range of positions which exist, see for example Matthew Paterson, 'International Justice and Global Warming', in Barry Holden (ed.), *The Ethical Dimensions of Global Change*, Macmillan, Basingstoke, 1996, pp. 181–201; Michael Grubb, James Sebenius, Antonio Magalhaes and Susan Subak, 'Sharing the Burden', in Irving M. Mintzer (ed.), *Confronting Climate Change*, Cambridge University Press, 1992, pp. 305–22.

The main alternatives to equal reductions discussed at the workshop were the following:

- *Responsibility-based differentiation:* here, commitments to abate climate change would be based on historical responsibility in contributing causally to climate change. This position was mentioned by several participants, and is a familiar part of the North–South debate regarding climate change. While this argument is partly reflected in the original differentiation in the Convention between Annex I and non-Annex I countries, it was suggested that it could be the basis for more subtle differentiation based on proportion of world emissions, and a possible future way for non-Annex I countries to also take on commitments. Some queried the desirability of translating responsibilities neatly into commitments. This approach becomes most relevant in the longer term when and if non-Annex I countries take on commitments on greenhouse gas emissions as a basis for broader differentiation, but no one at the workshop suggested it was a good basis for differentiating commitments among Annex I countries.
- *Rights-based approaches:* this was mentioned as a direct alternative to responsibility, or what some termed 'blame'-based approaches. Under this approach, commitments would be differentiated on the basis of the rights of each country (or individual) to emit greenhouse gases according (usually) to some measure of need. One means of putting this into practice is through the concept of 'minimum reasonable needs', mentioned in the presentation by Kawashima (Chapter 5).
- *Economic cost-driven criteria:* the focus of this approach from an equity point of view is to try in some way to equalize the economic burdens which states incur in meeting their emissions reductions. In discussions among OECD countries this is taken, sometimes implicitly, as the main implication of equity, and thus is likely to be the predominant basis for differentiating commitments among Annex I countries in the future. There are, however, different approaches to attempting this, and this argument is discussed further below.

A widespread view in the workshop was that these principled approaches will only be successfully put into practice if they meet crucial political tests. In other words, any differentiation of commitments will emerge from pragmatic considerations about what states will actually agree to. Finding agreements on commitments beyond what is contained in the Framework Convention might require negotiators to accept that some countries, often for domestic political reasons, cannot go beyond certain commitments, whereas others may be prepared to go further. Differentiation will proceed from this political feature of the climate negotiations, and formula-based approaches such as those mentioned above will have to be credible from this point of view in order to succeed. In an extreme version of this argument, one participant expressed the opinion that equity would play no role in developing future commitments, and that differentiation would proceed on a purely ad hoc basis.

As was shown by Greene's paper (Chapter 3 below), there is experience of this sort of pragmatic differentiation of commitments in other environmental agreements. The Protocol on Volatile Organic Compounds (VOCs) of the Convention on Long Range Transboundary Air Pollution, for example, establishes three types of commitment of differing degrees of stringency, from which states are obliged to choose only one.[4] Such a means of differentiation may allow negotiators to resolve the problem that if all states have to meet the same commitment, the level of stringency of that commitment will be determined by the state which is least able to reduce emissions (the 'lowest common denominator' effect). Some in the workshop suggested, on the other hand, that it may lead to a situation where a commitment is negotiated and differentiation allows states which feel unable or unwilling to meet that commitment to give themselves a weaker commitment. These participants argued that differentiation in practice would operate to reduce the overall effectiveness of future climate agreements. However, the discussion of Economies in Transition

---

[4] See Chapter 3, note 6; also Marc Levy, 'European Acid Rain: The Power of Tote Board Diplomacy', in Peter Haas, Robert Keohane and Marc Levy (eds), *Institutions for the Earth: Sources of Effective International Environmental Protection*, MIT Press, Cambridge MA, 1992, pp. 75–132.

(EITs) following the presentation by Sadowski (Chapter 6) suggested that this pragmatic form of differentiation could be used to encourage some countries, for example those EITs which have transformed their economies relatively successfully, to enter into commitments on the same basis as other Annex I countries.[5]

Lessons from experience

One way of evaluating the prospects for differentiation of commitments is to look at how it has been tried in other environmental agreements or international political fora. Greene's paper focused on other international environmental agreements. While he outlined a substantial number of ways in which states could pursue differentiation of commitments, the general direction of his argument was that, at least in the early stages of negotiations, such as those leading up to COP 3 in 1997, attempts to differentiate commitments to any great extent were unlikely to succeed. Discussions on Greene's paper focused on which approaches could work, or would definitely not work, in relation to climate change. Regarding the former, he mentioned that careful design of 'proxies' and the creative use of baseline dates and emissions levels could be used (see Chapter 3). But he emphasized the need for flexibility in using approaches developed for other agreements, as each particular problem has its own specific features and dynamics.

Additionally, the experience of the EU was mentioned throughout the workshop. The EU has tried to develop means of differentiating its commitment to stabilize CO_2 emissions by 2000 (both its unilateral target and its commitment under the Convention), but has found it particularly difficult to do so. Early on, in 1990, the EC had tried to differentiate its collective commitment through 'target sharing', whereby its overall target would be differentially distributed across the Community. This quickly

---

[5] The current commitments for EIT countries are of course the same as those for other Annex I countries. However, as already noted, the Convention allows for flexibility in meeting them. Sadowski makes clear that those within EIT countries regard their commitments as being different from those of other Annex I countries.

proved impossible to negotiate, and was considered not cost-effective, and against the principles of the internal market. The Community turned to an approach focused on policy measures, whereby the aim is to develop particular policy measures to share the effort of meeting an overall EC target across countries – a process in which the development needs of poorer members of the Community are taken into account ('burden-sharing'). Some at the workshop drew the lesson from this experience that the best way to differentiate commitments is to focus on policies and measures rather than targets. Others, however, suggested that the EU has found policies and measures (notably the energy/CO_2 tax proposal) impossible to agree, and therefore this approach is unlikely to be a useful model. The EU was also regarded by many as not being a useful model since it has a well-developed institutional structure for implementing and monitoring its internal agreements, unlike many broader international agreements.

Cost-driven criteria

Prominent in the debate on differentiation are arguments that commitments should be based on an attempt to equalize economic burdens among countries. A number of participants in the workshop mentioned such approaches. Equalizing economic burdens could be achieved through a number of means. One of these, discussed at the workshop and widely elsewhere, is of course Joint Implementation or tradeable permits. These, however, are not in themselves a means of differentiating commitments, but are alternative means of achieving economically efficient responses.

One way to use differentiation to meet these economic criteria is through macroeconomic modelling to estimate the costs to each country of reducing emissions, and to aim to equalize the marginal welfare loss. This approach was discussed in the workshop in relation to the MEGABARE model,[6] which has been used to analyse the different implications of a number of approaches to the question of equity for differentiation of commitments within Annex I countries. Different approaches to equity were

[6] Produced by the Australian Bureau of Agricultural and Resource Economics (ABARE).

translated into five operational rules for differentiating commitments: stabilization (equal reductions), population adjustment, horizontal equity (equal marginal reductions costs), polluter pays (such as the responsibility-based position outlined above), and tradeable permits.

For each of these rules, Annex I countries would have different commitments to reduce their CO_2 emissions. Each starting assumption produces a deviation from a basic aim of stabilization. These different schemes would have different welfare effects on Annex I (and non-Annex I) countries. The preference of those working with the MEGABARE model is for the horizontal equity approach, as the best way to differentiate commitments, although in the longer term, tradeable permits could be a more cost-efficient means of achieving global emissions reductions objectives. However, others at the workshop expressed scepticism that any specific economic models would ever gain international consensus as a basis for evaluating costs, given the well-known limitations of such models.

Some participants also suggested that national stabilization objectives would be an economically inefficient method of achieving overall goals. One cited a case study of Norway's options for reducing CO_2 emissions, and suggested that differentiation of targets would help reduce this inefficiency, and also that forms of emissions trading would help. The EU as a 'bubble' for commitments, expanding on the way it is already treated in the Convention, was cited as a possible model (see below for more on possible uses of groups of countries in relation to differentiation). Groups (including the EU) could agree to a target as a whole and differentiation of the target could be left to the states within that group.

The presentation by Kawashima detailed proposals to differentiate commitments according to various economic indices. She discussed how it would be possible to translate different indices relevant to CO_2 emissions into differential obligations on states to reduce their emissions. She started with a definition of minimum reasonable needs for CO_2 emissions, and suggested that the way to allocate objectives was on the basis of the difference between each state's present emissions and this long-term distribution of emissions (based on minimum reasonable needs). Then, when a short-term global target was set (such as a 10% reduction by 2010), this could be used to allocate specific reduction objectives to states. She then introduced four different ways of defining minimum reasonable need, and developed a model which showed by how much different industrialized countries would have to reduce emissions under each definition (see Chapter 5 below).

All of these approaches involve developing a formula for producing different commitments for different countries based either on one criterion (e.g. CO_2/GDP) or on a combination of criteria. Kawashima shows how each approach produces slightly different commitments for different industrialized countries if a 10% reduction target is adopted for those countries as a whole.

One of the fundamental problems with this sort of approach is that there are very many ways to construct these models and indices, and this often enables countries simply to advance their own preferred version through another, apparently more 'rational' means. Thus it would be clear to countries from the results outlined by Kawashima, for example, that they would benefit from advocating the use of one particular version of equity, or one particular index, over another. In addition, there are many sorts of disagreements, less clearly based in political interest, among modellers about how best to model abatement costs of emissions reductions, or other aspects of the models. As Greene highlighted in his presentation, it takes a long time within negotiations to secure agreement over formula-based approaches.

Regional differences and grouping proposals

As alternatives to formula-based discussions of differentiation, there was also much discussion of specific regional and national differences, and on ways of approaching differentiation from the point of view of grouping particular countries together, either to give each individual group a specific target or in a 'bubble' so that they have a collective target, leaving the group itself to decide how this target will be met.

One discussion of this approach concerned the EU. Building on the way the EU is already treated in the Convention, commitments could be assigned to the EU as a whole, which would leave it to the member states to negotiate ways of meeting that commitment. Some participants suggested this could be applied at the level of Annex I countries as a whole. However, a number of participants from outside the EU, particularly from Australia, expressed scepticism concerning this, on the grounds that many of the processes in the EU were opaque to them, and countries outside the 'bubble' may lack faith that those inside it would meet their commitments adequately. Many at the workshop referred to the problems that the EU – despite well-developed institutions for addressing collective concerns such as any common commitment on CO_2 emissions – has had in agreeing a strategy for its collective stabilization target under the Convention. They implied that the 'bubble' approach, if applied elsewhere, would experience these problems to a much greater extent. One participant suggested that this problem applies primarily because of the soft nature of the target contained in the FCCC; were this to be replaced by a legally binding target, such a 'bubble' approach might be more viable. Others questioned this, asking who would be legally responsible for ensuring the collective commitment was met in a 'bubble' approach, and how their responsibility would be transferred to member states.

Other proposals for grouping were based on other ways in which countries might be grouped together to take on the same commitments within the group, but with each implemented individually and differentiated from the commitments of other groups. One question was whether some groups of countries might begin to take on the commitments currently adopted by Annex I countries. The presentations by Sadowski and Lee both suggested such possible groupings. Sadowski discussed how some of the EIT countries might be able to sign up to the Annex I commitments without the qualifications contained in their existing commitment to limit their emissions, given their economic performance and success in achieving economic transition. Some participants agreed with this possible grouping, but disagreed over the countries which might be included. Lee also suggested that some non-Annex I countries, the economically advanced developing countries, might be able to sign up to commitments in the relatively near future, if these were couched in an appropriate form (see more below on non-Annex I countries).

Sectoral disaggregation

The workshop discussed an additional question relevant to differentiation: that focusing on differences between sectors may make negotiations more manageable in general, as well as specifically in relation to differentiation. Different sectors clearly exhibit different characteristics in relation to the dynamics of their greenhouse gas emissions and other features (mentioned below). Therefore disaggregating among them may make reaching agreement easier by breaking the negotiations down into more manageable chunks. It also might allow negotiators to highlight those sectors which matter most in addressing climate change (e.g. by focusing on those with the fastest emissions growth rates).

The paper by Grubb (Chapter 4) focused on how, when different economic sectors which emit CO_2 are examined, significant differences in the characteristics of those sectors suggest alternative ways of differentiating commitments. Grubb argued that any solution to the differentiation problem must ultimately be based on an understanding of these different sectoral characteristics and driving forces. The positions of and concerns raised by countries largely reflect differences in their economic structure. Any credible approach to differentiation must ultimately reflect these underlying factors.

Grubb suggested that this would have advantages beyond those of differentiation, because it would enable the structure of agreements to be varied to reflect differing sectoral characteristics relating to monitorability, the internationalization of some industries, precision and ease of control, etc. He suggested that economic sectors could be split usefully into three. Agriculture could be treated separately, as Sweden has already proposed in the negotiations. Energy-related emissions could be split into industrial (including electricity production) and service-related emissions on the one hand, and buildings and transport-related emissions on the other. Grubb suggested that these two had significantly different features, relating to their underlying driving forces and how they could be indexed; their differing trends; the different ways in which they affect government legitimacy (for the former through effects on competitiveness, for the latter through electoral concerns); the different institutional structures needed to implement reduction measures for them; their different fuel mixes; and their different relationship to country-by-country differentiation of targets. Another participant agreed with Grubb that sectoral differences are crucial for considering differentiation of commitments, but suggested another way of dividing sectors, arguing that, at least as far as underlying emissions trends were concerned, the most significant sectoral differences were between electricity emissions, transport emissions, and others. Electricity and transport are the only sectors where emissions have grown consistently in OECD countries since the mid-1970s, and have significantly greater rigidities than other sectors, notably in relation to price elasticity. Grubb agreed that transport could readily be considered separately, but maintained that electricity and industrial emissions would ultimately need to be treated together for future commitments because of their converging characteristics and substitutability.

However, participants voiced concerns about using differences between sectors as a basis for differentiation of commitments. One point raised was that it could reduce the flexibility which countries had in meeting commitments, since these would be sector-specific and would therefore limit the capacity to choose where to aim to meet reduction objectives. A representative of an environmental group argued that if commitments concerning industry were separated, then these commitments could be compromised further, since industry is particularly well organized politically at global negotiations on climate change.

Non-Annex I countries

Most of the discussion in the workshop related to the possible differentiation of commitments within Annex I countries, reflecting the fact that only those countries will have commitments to limit their emissions at least until COP 3 in 1997. However, some papers and other parts of the discussion opened a debate on possible differentiation of commitments within non-Annex I countries. It was suggested that basing commitments clearly on responsibility for causing climate change would mean that in the long term at least some commitments by developing countries would become necessary, as they began to be significant contributors to climate change; and suggested that non-Annex I countries could gradually take on commitments to reduce the rate of growth of their emissions. The paper by Lee (Chapter 7) also discussed this, focusing on those developing countries who are economically advanced, and therefore perhaps economically better capable of taking on commitments to limit emissions. Lee suggested three possible courses of action when considering additional commitments for advanced developing countries such as Korea: OECD income parity, where those countries would be obliged to undertake commitments once their per capita incomes reached the OECD average; a commitment to energy efficiency improvement; and/or the removal of subsidies on energy. There was considerable concern at the workshop about the income parity proposal in particular, since it would seem to downplay the necessity of a global response to climate change.

Malpede (see Chapter 8) also made similar suggestions and proposed that some non-Annex I countries might be able to take on commitments in the future. He suggested some criteria by which those countries might be specified, such as being close to the fulfilment of criteria for joining the OECD, GDP, total emissions, GDP and emissions per person, human development index, or carbon efficiency. He then outlined different ways in which they could be involved in commitments: on the basis of voluntary involvement, as allowed for in Article 4.2g of the Framework Convention; or on the basis of legal obligation, although this is unlikely, at least in the short term, since the Berlin Mandate does not allow for new commitments on non-Annex I countries. He also mentioned the possibility of linking acceptance of commitments to mitigate climate change to accession to the OECD, but had reservations about the conditionality involved, and referred to the idea originally put forward by Sebenius, that a coalition of Annex I countries might gradually begin to include some major non-Annex I countries such as China, India, Brazil or Indonesia.[7]

By contrast many participants, in particular those from developing countries, suggested that such discussion was premature, since it encouraged a sense that it was reasonable for developing countries to take on emissions limitation commitments in the near future, which most develop-

---

[7] See James K. Sebenius, 'Towards a Winning Climate Coalition', in Irving Mintzer and J. Amber Leonard (eds), *Negotiating Climate Change*, Cambridge University Press, Cambridge, 1994, pp. 277–320.

ing countries do not accept. Malpede, however, was explicitly critical of this negative attitude. He emphasized that many developing countries were adopting this posture in formal negotiations, but in bilateral relations with industrialized countries were engaging in many particular arrangements to limit greenhouse gas emissions. He called this the 'double speech syndrome', and suggested that the contradiction was not helpful in developing responses to climate change.

Differentiation and flexibility

One general theme which came up in the workshop was the distinction between differentiation and flexibility. Differentiation of commitments is one way of providing flexibility for countries in negotiating their commitments. However, one argument was that in practice flexibility might be just as easily achieved without having to go through the difficulties in negotiating differentiated commitments. This could be done by allowing countries flexibility in meeting their commitments, even where those commitments might be undifferentiated. The distinction outlined by Kinley and Terrill in Chapter 2 between concrete and contextual differentiation is another way of phrasing this; the latter is focused on allowing flexibility for countries in meeting their commitments, and building on the many qualifications already contained in the Convention which operate in this way.

Such arguments for flexibility are already well known in the climate debate in relation to arguments for tradeable permits systems and for Joint Implementation. Both of these schemes allow states flexibility in meeting their commitments, by allowing them to invest overseas (either directly or through the purchase of surplus emissions permits from other countries) to meet their commitments. In addition, the 'comprehensive approach', where all greenhouse gas sources and sinks are included as a basket, would increase the flexibility of countries in meeting their commitments.[8]

---

[8] There are of course notable problems with the comprehensive approach, particularly concerning the construction of an index which equalizes the contributions of each gas to climate change (where the choice of timescale has important effects on the relative contribution of each gas) and the fact that greenhouse gas emissions and uptake from sinks are difficult to monitor accurately except for fossil fuel CO_2. See Michael Grubb, David Victor and Chris Hope, 'Pragmatics in the Greenhouse', *Nature*, 354, pp. 348–50, December 1991.

However, as mentioned above, Greene showed how other agreements allow for flexibility in other ways. As negotiations proceed on how to implement or operationalize particular commitments, they get progressively more technical, and in these technical debates there is often much room for manoeuvre in using them creatively to increase the flexibility which countries have in meeting their commitments.

Conclusions

As outlined above, the workshop at Chatham House discussed in detail a number of important issues related to the question of differentiation of commitments. A broad conclusion, although not one shared by all participants, was that in the near future flat-rate reductions would remain the default option. Several participants expressed concern about and objections to this way of proceeding. However, as yet there have been no proposals for credible specific alternatives which could be negotiated by COP 3. What is possible is some small deviations from flat-rate reductions, such as qualifications to commitments for some countries already contained in the Convention, and perhaps some more developments in flexibility such as could be provided by emissions trading or more extensive use of soft law in supplementing the more formal commitments contained in any legal instrument. Substantial proposals for differentiating commitments are unlikely before COP 3 in 1997. However, looking further into the future, attempts to devise ways to achieve this look increasingly likely, as flat-rate reductions beyond those currently envisaged become politically considerably more difficult to negotiate.

Chapter 2

Differentiation and the FCCC: provisions, proposals by parties, and some options

*Richard Kinley and Greg Terrill**

Possible differentiation among Annex I Parties is an important issue in the Ad Hoc Group on the Berlin Mandate (AGBM) – and one on which there has not been enough discussion to date.

The secretariat was asked by the AGBM Bureau to prepare a paper for the July 1996 session of the AGBM on possible indicators for differentiation among Annex I Parties. We were also requested to do a review of existing international agreements to see what might be relevant to the work of the AGBM – including with regard to differentiation.

In developing these two papers we expanded a concept originating in the academic literature to differentiate differentiation. Two general kinds of differentiation have been identified – 'concrete' and 'contextual'. By concrete differentiation we have understood that different Parties, or different groups of Parties, have different commitments. Contextual differentiation, on the other hand, sees all relevant Parties having the same commitments – but the achievement of such commitments would depend upon, or take account of, national contexts.

The Framework Convention on Climate Change reflects differentiation in a number of ways. First, there is a clear differentiation between Annex I and non-Annex I Parties, and between Annex II and other Parties. These are examples of concrete differentiation. The FCCC has a number of other provisions, reflecting mostly contextual differentiation, and we have the negotiators of the FCCC to thank for the list that follows.

There is some differentiation among Annex I Parties. First, and most important, Article 4.6 of the Convention provides that a certain degree of flexibility shall be allowed by the Conference of the Parties to the Annex I

* The views expressed are personal and should not be construed as representing the position of the Convention Parties or of the UNFCCC secretariat.

Parties undergoing the process of transition to a market economy. These Parties are identified by an asterisk in Annex I to the Convention.

Second, Article 3.1 refers to 'common but differentiated responsibilities and respective capabilities', and Article 3.2 refers to 'those Parties... that would have to bear a disproportionate or abnormal burden under the Convention'.

Article 4.2(a) refers to Annex I Party commitments while also 'taking into account the differences in these Parties' starting points and approaches, economic structures and resource bases, the need to maintain strong and sustainable economic growth, available technologies and other individual circumstances, as well as the need for equitable and appropriate contributions by each of these Parties to the global effort regarding that objective'. Article 4.2(a) is also reflected in paragraph 2(a) of the Berlin Mandate (decision 1/CP.1).

Articles 4.1, 4.3 and 4.10 also reflect some degree of differentiation *vis-à-vis* the implementation of commitments by Annex I Parties, employing the following concepts:

- national and regional development priorities, objectives and circumstances;
- burden-sharing;
- taking account of economies that are highly dependent on income generated from the production, processing and export and/or consumption of fossil fuels and associated energy-intensive products and/or the use of fossil fuels, in cases where such Parties have serious difficulties in switching to alternatives.

The Convention also differentiates among non-Annex I Parties, making mention of groups of Parties including:

- least developed country Parties;
- small island developing country Parties;
- developing country Parties that are particularly vulnerable to the adverse effects of climate change;
- economies which are highly dependent on income generated from the

production, processing and export, and/or on consumption of fossil fuels and associated energy-intensive products.

Such a list makes clear that, whether or not Parties determine to pursue differentiation any further, it is already a part of the Convention. The limited discussions in the AGBM to date would suggest that there are two general approaches to the question of whether there should be differentiation among Annex I Parties in the protocol or whether another legal instrument should be adopted at COP 3. Some Parties have spoken in favour of a 'flat-rate reduction' objective, because of the difficulty they perceive in negotiating a differentiated regime. They also see such an approach as equitable, by virtue of the fact that progress is measured against a Party's own national emissions in the base year.

Other Parties see deficiencies in the 'flat-rate' approach and have advocated a differentiated approach. In their view, this would better respond to differing national circumstances, particularly with regard to costs of abatement and levels of economic development and growth. They also see differentiation as more equitable and efficient and argue that it would enhance the cost-effectiveness of the emissions reduction effort.

In the second and third meetings of the AGBM, the discussion of differentiation has taken place under the agenda item on quantified emission limitation and reduction objectives (QELROs). Differentiation with regard to policies and measures could also emerge as an issue. The discussions to date have been very limited and rather abstract. Only one concrete proposal has been offered – that of the Russian Federation. No substantive conclusions on differentiation have emerged.

In their submissions, some Parties have suggested a variety of relevant considerations. These fall into three categories, covering concepts relevant to differentiation, specific indicators for differentiation, and approaches to differentiation. A few examples from each category follow.

First, the concepts of equitable burden-sharing, and the fair distribution of costs, have been advanced by several Parties. A number of approaches have been put forward, including the following four:

- countries in a given category or across categories could choose to form clusters, combining their emission reduction objectives and sharing the costs and benefits of achieving them;
- there could be a protocol with different provisions for different groups of countries;
- a tax on CO_2 could be adopted;
- countries with relatively high domestic costs of measures might do more internationally and countries with relatively low domestic costs would do more at home.

Second, the following possible indicators for differentiation, which might be used either individually or in combination, have been specifically mentioned:

- emissions per square kilometre of a country's territory basis;
- availability of sinks;
- 'critical economic loads';
- levels of production and consumption of energy per capita;
- per capita emissions;
- per gross domestic product (GDP) or gross national product (GNP) emissions;
- share of global emissions;
- share of the respective Party to global warming;
- carbon intensity of primary energy use;
- marginal costs of abatement (per unit of emission reduction);
- levels of production and consumption of energy per capita.

This lengthy list demonstrates again the breadth of options that Parties have in determining how they might wish to pursue differentiation.

Third, if Parties wish to consider differentiation further, they will need to address several basic approaches. Would differentiation be 'concrete', as discussed earlier, whereby different Parties or different groups of Parties would have different commitments? For example, Parties could be divided into two or more groups, with each group having different QELROs or different base or target years.

Or would the differentiation be 'contextual', in that Annex I Parties would share the same commitment or commitments, but the achievement of such commitments would depend upon, or take account of, the national contexts?

One approach is that differentiation could be based upon indicators. For this possibility, the secretariat has identified three basic clusters of indicators, based upon national emissions, national circumstances, and costs of action.

Differentiation could also be based upon Parties 'self-selecting' into one of several 'commitment categories' that the AGBM had established. For example, the AGBM could adopt two or more QELROs and invite Parties to subscribe to the commitment which was most suited to their national circumstances. In this way, different groups of Parties would be constituted. Parties could be invited to outline the reasons for their choice, to help clarify those factors most relevant to differentiation. Alternatively, placement in such commitment categories could be negotiated.

In another approach to differentiation, the AGBM could decide to allocate commitments based upon complex economic models or differentiation formulae which, for example, attempted to define a set of actions that would minimize the overall international cost of abatement.

There could also be some combination of, or alternatives to, these options.

There are a number of factors which might potentially be considered relevant to differentiation:

- national circumstances, such as physical characteristics, demographic characteristics, energy profile, or socio-economic characteristics;
- cost, and capacity to meet costs, such as indicated by, for example, national GDP;
- 'equalizing the costs of mitigation', or emissions abatement efforts, across countries; 'net national economic cost', measured by GDP forgone, absolute costs or some other formula; or some other means of equating the economic load of abatement efforts;
- some other method.

At this stage, there is no consensus on any of these basic issues.

At COP 3, in December 1997, a protocol or other legal instrument is to be adopted to strengthen the commitments applying to Annex I Parties. Before then, there are three negotiating sessions:

- AGBM 5 in December 1996;
- AGBM 6 in March 1997;
- AGBM 7 in summer 1997.

AGBM 4 in July 1996 was the occasion for the first in-depth discussion of the issue. Thus, time is short. There is no consensus yet in favour of a differentiation regime. Few concrete proposals have been presented by Parties and none has generated a critical mass of support. Furthermore, the issues involved are complex and highly political. In these circumstances, the onus is upon Parties to come forward with realistic, specific proposals for differentiation, or to decide not to pursue differentiation further in this round, and to concentrate efforts in other areas.

Chapter 3

Lessons from other international environmental agreements

Owen Greene

Introduction

There are now hundreds of international environmental agreements, of which at least 120 are multilateral with contemporary relevance and some international legal substance. The task of identifying lessons from these for the development of commitments in the Framework Climate Change Convention needs to be approached with caution. Reliable generalization is difficult. Moreover, no agreement should be directly modelled on another. The effectiveness of a particular approach depends greatly on many specific factors, including: the nature and scope of the issue area; the numbers and characteristics of states and key stakeholders involved, and their patterns of interests and concerns; the state of knowledge about the problem and about possible solutions; and the international context. There is wide variation in the designs of previous international agreements, and opinions differ about how effective each turned out to be.

Nevertheless, experience with other international agreements does appear to offer some important or suggestive lessons for debates about the possible differentiation of commitments under the FCCC. Most agreements have had to cope with the fact that the burdens and benefits of implementing 'equal' commitments will be unevenly distributed among Parties. Similarly, it is typical that negotiations involve countries that differ in what they are able or prepared to do to tackle the problem under consideration. At any given time, some governments have always valued a particular environmental good more highly than others, or given relatively high priority to achieving effective international environmental protection.

These facts of international life make the possibility of differentiated commitments at least a background issue for most multilateral negotiations. Moreover, a search for economically efficient 'least global cost' strategies for tackling global environmental problems also implies exploring differential treaty obligations.

However, the central problem for the design of international environmental commitments is their negotiability, and their effectiveness in actually promoting desired changes in behaviour. The next section identifies what seem to me to be some important and reliable lessons from experience with other international agreements relevant to whether and how to try to negotiate differentiated commitments under the FCCC. The following section then outlines ways in which agreements involving 'equal' commitments have been able to address problems of unequal burdens and differences in what governments are prepared to do. Finally, some conclusions are drawn for the negotiation of strengthened commitments under the Climate Change Convention.

Relevant lessons for the negotiation of differentiated commitments

On the basis of a review of the range of experience with international environmental (and other) agreements, and bearing in mind the cautionary remarks made above, the following generalizations appear to me to be valid and reasonably reliable.[1]

---

[1] The generalizations are based on the author's research and reading of the literature on the negotiation and development of a range of international environmental agreements (and also arms control agreements); and also the findings of some reviews by other researchers. Among the latter, the following were particularly useful: P. Sand, *Lessons Learned in Global Environmental Governance*, World Resources Institute, Washington DC, June 1990; E. Parson and R. Zeckhauser, 'Equal measures or fair burdens: negotiating environmental treaties in an unequal world', in H. Lee (ed.), *Shaping National Responses to Climate Change: A Post-Rio Guide*, Island Press, Washington DC, 1995, pp. 81–114; O. Young, *International Governance: Protecting the Environment in a Stateless World*, Cornell University Press, London, 1994; I. Zartman, 'Lessons for analysis and practice', in G. Sjostedt (ed.), *International Environmental Negotiation*, Sage/IIASA, London, 1993, pp. 262–74; L. Susskind, *Environmental Diplomacy: Negotiating More Effective Global Agreements*, Oxford University Press, Oxford, 1994.

Allocative efficiency has rarely been a key determinant of commitments
From a global perspective, there is a clear case for preferring treaty commitments that tackle the environmental problem adequately at minimum overall cost. This would imply allocating obligations to reduce emissions of pollutants, for example, so that the greatest emissions reductions would take place in countries where they could be achieved most cheaply. An ideal set of differentiated commitments from this perspective would allocate responsibility for emissions reductions so that the marginal costs of environmental protection measures would be equalized across countries and regions.

However, such 'allocative efficiency' principles have rarely been a primary or dominant factor in the design of successfully negotiated environmental commitments. In part this is probably due to great uncertainty about what the minimum cost allocations or marginal costs would be. Even in the rare event that economists can more or less agree on such assessments, representatives of the different political and cultural communities involved in a multilateral negotiation will typically disagree profoundly on how to value different factors. Moreover, the implications of such an approach tend to be politically unacceptable among international negotiating partners. Since marginal abatement costs tend to vary greatly between countries, it implies large differences in national commitments.

Negotiators know that this means that agreements designed according to such principles would be relatively non-negotiable. In practice, therefore, they have preferred to focus simply on finding commitments that all key parties can accept and recognize as useful. Where these can be designed so that they are also reasonably economically efficient overall, all well and good, but this has typically been a secondary factor in shaping successfully negotiated international obligations.

'Problem-solving' negotiating approaches are relatively successful
Questions relating to the distribution of costs and benefits are important factors in all substantial international negotiations. However, distributional issues do not always need to dominate the negotiation process. In practice, life in international environmental negotiations is rarely at the 'Pareto

frontier', where negotiations must be about the distribution of fixed payoffs rather than about finding ways to achieve the production of expanded benefits.[2] Experience shows that a negotiating framework which emphasizes integrative bargaining, or problem-solving processes focused on searching for 'positive sum' solutions to shared problems, is more likely to succeed than a framework in which distributional debates dominate.

In this context, emphasizing differentiated commitments risks stimulating distributional arguments more than is necessary. In principle, if there is broad agreement from the outset that differentiated commitments of some sort are justified and necessary, there is no contradiction between negotiating for such commitments and adopting a 'common problem-solving' or 'integrative' approach. However, without such initial agreement, early proposals emphasizing differentiated obligations can increase the risk that negotiations become bogged down in distributional wrangling.

'Defensible equity' is normally vital for effective agreements

One of the motivations for proposing differentiated commitments is to make burdens more equitable. Perceptions of fairness or equity are key factors in the negotiation, implementation and development of international environmental agreements. States cannot legally be obliged to join agreements that their governments or peoples believe to be deeply unfair. Moreover, most decisions are taken by consensus in environmental regimes. While political and other pressures may be informally applied, the stakes have rarely been high enough for powerful countries to try to force an unwilling government to sign what it perceives to be an unjust environmental agreement. In any case, governments need to persuade their legislatures that an agreement is acceptable in order to achieve treaty ratification. Furthermore, governments and their peoples tend to neglect implementation of obligations which they perceive to be grossly unfair, even if they are formally bound by them.

---

[2] See, for example, S. Krasner, 'Global communications and national power: life on the Pareto frontier', *World Politics*, 43 (April 1991), pp. 336–66, and O. Young's critique of the applicability of this approach (O. Young, op. cit., chapter 5).

At first sight, awareness of the importance of fairness might be taken to imply that multilateral commitments should be negotiated according to clear and agreed principles and criteria for equity. However, experience shows that in the development of international environmental agreements, it is rare for perceptions of equity to be based mainly on principles of equity of burden. Typically, many notions of equity are in play, involving, for example, equal percentage changes from the status quo; equal end results; equal (per capita or per state) quotas for pollution or resource extraction; burdens distributed according the 'polluter pays' principle or historical responsibility or capacity to pay; 'first come, first served'; respect for 'basic rights', historical traditions or cultural identity; inter-generational equity; and so on.

Precisely because it is vital that agreements are widely perceived to be acceptably fair, equity arguments need to be handled with care. In any complex society, one can be sure that different communities will have different ideas about what is fair, reflecting different world views, values and ideologies as well as differing interests. This applies *a fortiori* to the range of parties and stakeholders involved in multilateral or global agreements. Attempts to develop multilateral obligations based on a single conception of equity have usually failed. Whatever compromises may be made in the fine print, governments and other key actors need to perceive that they have not conceded basic points of principle or core values, or accepted an agreement that is basically 'unfair'. Moreover, they need to be able to defend themselves against accusations that they have done so when they present the agreement domestically for ratification and implementation.

Thus, successful environmental agreements have been 'defensibly equitable' from a wide range of standpoints. Inevitably, they have not been 'fair' according to every equity criterion. However, given that in most countries a range of different, and potentially contradictory, notions of equity typically coexist and command potential support, this has not been necessary. What has been required is that the range of different equity criteria that have been accommodated has been sufficient to provide scope for the development of 'winning coalitions' in support of the agreement, both internationally and domestically.

However, perceptions of fairness arise from a political process. If a government consistently focuses on only one criterion for fairness, for example equity of burden, it increases the likelihood that any final international agreement will be judged domestically according to that criterion above all others, and may be found lacking. Although reducing domestic room for manoeuvre is a well-tried negotiating tactic, it is a game that everyone can play. Moving forward with multiple, and slightly vague, notions of equity has tended to be a more politically robust approach, providing welcome flexibility in the difficult task of finding mutually beneficial international agreements.

Equality of burden has not been a dominant factor in the design of international environmental commitments

In many other environmental negotiations, reasonable arguments could be (and were) made in favour of differentiated commitments to equalize burdens between parties. However, as indicated above, equality of burden of implementation has almost always been only one of several equity principles in play in the negotiations, and it has rarely been dominant in the formulation of commitments.

There are a number of reasons for this, including the fact that calls for equality of burden have rarely been seen by most participants to command particular moral authority in addressing environmental issues. Moreover, there have typically been major differences in countries' levels of concern and in their willingness to take measures to tackle the problem under consideration.

Furthermore, as with cost-efficiency, equality of burden has normally been an unpromising basis for successful negotiations. It encourages an immediate focus on distribution issues. The uncertainty and complexity of assessments of overall costs and benefits mean that negotiations are vulnerable to becoming bogged down in special pleading. The negotiating focus encourages governments to find arguments to show that proposed measures are particularly difficult or burdensome for their country, instead of looking for what they *can* achieve at reasonable cost. In multilateral negotiations with many participants, this tends to lead to protracted negotiations and 'least common denominator' commitments.

Such negotiations can be facilitated by focusing on 'depersonalized' formulae designed to determine commitments that aim to take into account objective factors affecting costs of implementation. Unfortunately there are normally several such factors. This leads to complex, multi-criteria formulae, with contestable weightings for each factor. In practice, these have not been a recipe for successful negotiations and effective agreements, as discussed below.

Simply expressed commitments involving equal measures are usually more negotiable, particularly in the early stages of environmental regime development

Most international environmental agreements have involved simply expressed and relatively undifferentiated commitments. There has been a clear preference for bans; equal allocations or entitlements; equal percentage cuts; common standards; or equal targets. Where basic obligations have been differentiated, this has tended to be according to a very few basic classes of states. For example, the Montreal Protocol and the Biodiversity Convention, as well as the FCCC, have essentially divided parties into developed and developing countries, with industrialized countries having more stringent obligations.

The tendency to simple, equal agreements has been widely noted as a phenomenon that requires explanation. In principle, complex and differentiated commitments would appear to offer more opportunities to find agreements that can both address the environmental problem in a reasonably efficient way and accommodate the wide variety of interests, values, priorities and circumstances of the parties and stakeholders involved. Moreover, *national* regulations relating to the environment and other areas often successfully involve complex multi-criteria formulae and differentiated obligations.

Explanations for the strong tendency towards simple, equal commitments in *international* agreements focus on the distinct characteristics of multilateral bargaining.[3] Within functioning states, there is a clear

---

[3] See, for example, discussion in Parson and Zeckhauser, op. cit.

governmental structure for taking decisions, imposing solutions, and developing and maintaining complex packages and trade-offs between issues and across time. This helps states to overcome domestically some of the main negotiability problems that are posed by complex and differentiated packages of regulations.

In international, and particularly global, environmental negotiations, there is typically no such supervening authority. International institutional resources and structures are usually relatively weak, and restricted to particular issue areas. Politically, international society is relatively fragmented.

In this context, there are several strong advantages to focusing on negotiating simple proposals for commitments involving equal measures. Simplicity facilitates clear communications among the many actors involved in multilateral negotiations. By accepting simplicity as a constraint, negotiators greatly reduce the number of proposals that could potentially be considered, and thus make the negotiating process more manageable. Moreover, focusing on deals involving equal measures imposes some constraint on opportunist or extreme self-seeking proposals: all parties would at least be obliged to bear some costs as well as gain the environmental benefits of collective actions.

Furthermore, simply expressed equal measures have a political and psychological salience that more complex or differentiated commitments lack. Arguably, for example, a ban or equal cuts represent prominent focal points from which any departure is perceived to be a big change. This enhances the stability of the bargain, because it helps to deter negotiators from attempting to change the deal to secure marginal advantages. Commitments to equal measures also have an immediate appearance of fairness, and thus are well on their way to meeting the 'defensible equity' requirement discussed above. Complex and differentiated bargains tend to be more vulnerable to charges of unfairness and more likely to be picked apart, even if in fact they accommodate each country's interests more fully or share the burden more equally.

Nevertheless, there are some examples of multilateral agreements where obligations are calculated by complex formulae which take into account a range of different factors, and which involve differentiated commitments. This is the case, for example, for some regional fishing agreements allocating quotas and regulating fishing practices. Similarly, the EC's Large Combustion Plants (LCP) directive in 1988 specified highly differentiated obligations on member states to limit their sulphur and nitrogen emissions. In 1994, parties to the Convention on Long Range Transboundary Air Pollution (LRTAP) to limit acid rain agreed to regulate such emissions according to the 'critical loads' concept, which implied unequal national limits.

However, it is important to recognize the particular contexts in which such complex and differentiated commitments were negotiated. Fishing agreements are examples of a particular class of environmental problems (relating to the management of 'common pool' resources), where agreement must be achieved among a restricted group of 'users' to manage a limited common resource sustainably. Because it is much the same group that directly suffers the consequences if effective agreement is not achieved, there is reason to think that there is a relatively good chance of users being able to agree effective cooperative actions in such cases.[4] Nevertheless, it must be noted that the record of success has not been good so far. In spite of the complex formulae and shared interests, agreement has mostly been achieved only by setting overall fishing quotas so high that fish stocks have continued to decline. Only in the mid-1990s were stringent reductions in quotas established in some areas, such as in northwest European waters, and this was in the context of potentially catastrophic reductions in fish populations and strong action through EU institutions.

Negotiations for differentiated environmental commitments among EU member states, such as the LCP directive, were similarly achieved by making full use of the decision-making procedures and institutional resources and linkages of the EU. Moreover, they were facilitated by the relatively intense and long-term interactions among key stakeholders in western Europe. The EU has many of the systems traditionally available only to states to help it overcome the obstacles to achieving complex and differ-

---

[4] See, for example E. Ostrom, *Governing the Commons*, Cambridge University Press, Cambridge, 1990; and also discussion in A. Weale, *The New Politics of Pollution*, Manchester University Press, Manchester, 1992, chapter 7.

entiated environmental bargains. Even so, it took five years of twice-weekly meetings to agree the LCP directive.[5]

As already noted, there are agreements such as the Montreal Protocol, the Biodiversity Convention and the FCCC, where commitments have differentiated between industrialized and developing countries. However, little differentiation of national commitments *within* either of these two classes of states has been agreed.[6] LRTAP provides perhaps the strongest precedent for achieving differentiated international commitments between developed states by multilateral negotiations, through its adoption of the 'critical loads' approach to allocating emission limits. However, LRTAP itself adopted the simple, equal measures approach in its first years of development. In fact, it took years of negotiation to establish LRTAP in 1979 and still longer to negotiate commitments to equal 30% cuts in national sulphur emissions (agreed in 1985) and to freeze NOx emissions (agreed in 1988). It was only after several years of implementing, reviewing and developing these commitments that it proved possible to adopt the critical loads approach in 1994.[7] During those years the regime became established. By the early 1990s, the parties and stakeholders had become familiar with one another, with the RAINS model of acidification[8] (which is central to the setting of obligations in the critical loads approach), and with the costs of implementation.

The main lesson from LRTAP in this context seems to be that it is possible to agree complex differentiated commitments in an international environmental regime, but only after the regime has become well estab-

---

[5] See, for example, M. Grubb, *The Greenhouse Effect: Negotiating Targets*, RIIA, London, 1989; M. Levy, 'International Cooperation to Combat Acid Rain', in *Green Globe Yearbook 1995*, pp. 59–68.

[6] Although there has been strong awareness of the differences in the circumstances of OECD states and post-communist 'countries with economies in transition', and of least developed states and other developing countries, this has only been reflected in differential environmental commitments to a very limited extent.

[7] M. Levy, private communication, 1996. Note also that LRTAP's 1991 VOCs Protocol, which focused on 30% emissions reduction commitments for most parties, allowed a class of 'light emitters' to only freeze their emissions, thus making the agreement acceptable to states such as Bulgaria, Greece and Hungary

[8] Developed by the International Institute of Applied Systems Analysis (IIASA) in Laxenburg, Austria.

lished, detailed common understandings have developed of the problem and of responses and simple, equal commitments have been agreed and effectively implemented.

Ways of designing commitments to address differences in burdens and interests

Experience from other international environmental agreements discussed above indicates that differentiated commitments have only rarely been agreed, particularly outside the EU. Simply expressed, equal measures have generally been found to be more negotiable. Similarly, allocative efficiency or equity of burden have rarely been essential for successful environmental agreements. Nevertheless, such simple, 'equal' agreements have had to find ways of addressing concerns about inequitable burdens and coping with the fact that some states are prepared to do more than others. This section outlines ways in which this has been done.

Astute choice of 'equal' measures

Even when negotiations have focused on agreeing undifferentiated commitments, there have normally been several useful types of measures available to choose from. Each type of measure will have different implications which could greatly affect negotiability and the likely effectiveness of the agreement. In particular, different types of measures will distribute the burdens and benefits differently among parties and stakeholders. Astute choices of which measures to negotiate have enabled negotiators to accommodate parties' varying concerns and interests sufficiently for useful environmental agreements to be achieved.

For example, 'equal' commitments to reduce emissions of pollutants can take the form of equal percentage reductions compared with a baseline year; bans on sub-sets of pollutants; equal emissions entitlements; equal absolute emissions reductions; equal taxes on emissions; bans on (*de facto*) emissions subsidies; equal controls on 'proxies' (such as fuel types, or consumption of the inputs to the polluting activities or of their products); equal ambient concentrations of pollutants; equal measures to tackle impacts; equal taxes or bans on specified polluting technologies or practices; adoption of equivalent technical standards; equal trade restrictions; or participation in international information-exchange, education or research programmes to promote best practice. Moreover, emissions entitlements or reductions, for example, can be measured in many ways, such as nationally, per head of population, or per unit of GNP. Controls can be applied separately to individual pollutants or to groups of pollutants.

In this way, many different types of equal measures become potentially available for negotiation. Each of them can be expressed simply, and can in principle stand a good chance of being widely perceived as uncontrived and defensibly equitable. In practice, some will imply a much more acceptable overall balance of burdens and benefits than others; and the process by which one such measure gets singled out as a focus for negotiation is complex and often highly contingent. Nevertheless, it provides one of the key ways in which chairs of the various bodies of the Convention, influential policy-entrepreneurs and 'lead' states can facilitate effective agreements, by identifying and promoting types of measures which all key actors are likely to find acceptable and useful.

Creative definitions

Once negotiations have begun to focus on a particular type of measure, there is normally further scope to accommodate different interests and concerns by careful, and sometimes imaginative, formulation of definitions. For example, once it had been agreed in the Conventional Forces in Europe negotiations in the late 1980s that NATO and the Warsaw Pact should reduce their tanks in Europe to equal numbers on each side, negotiators promptly settled down to decide how to define a 'tank' in a way that accommodated both sides' key concerns. Similarly, the formulation of definitions of 'consumption' of ozone-depleting substances in the Montreal Protocol, or a 'critical load' in LRTAP, or 'commercial whaling' in the International Whaling Convention, has played an important role in the negotiation of these environmental agreements, partly because of the implications for the distribution of costs and benefits.

Choices of baselines and target dates

Many environmental and other agreements have defined commitments in terms of relation to changes from a 'baseline'. For example, the Montreal Protocol specified limits on annual consumption of CFCs in comparison with a 1986 baseline, and the 1985 Sulphur Protocol of LRTAP required 30% reductions in national sulphur emissions by 1994, compared with a 1980 baseline. The actual commitment for each state depends on the level of its controlled activities during the baseline year. The choice of baseline year can thus be an important factor in determining not only the overall stringency of any commitment but also the distribution of burdens. For example, the change in stringency of environmental commitments implied by moving the baseline year from 1991 to 1988 will typically be much more dramatic for former Warsaw Pact countries than for OECD states. Choice of baseline year therefore becomes an additional instrument for making an 'equal measure' agreement more widely acceptable.

In fact, some previous environmental agreements have made more extensive use of flexible or negotiated baselines for this purpose. Within LRTAP, for example, the 1991 VOC Protocol required parties to reduce their VOCs emissions by 30% but allowed them to choose their own baseline year between 1984 and 1990 (with 1988 as a default baseline). Similarly, some of the limits on consumption of ozone-depleting substances in the Montreal Protocol involve nominal 'calculated' baselines (for HCFC or methyl bromide, for example), or baseline years set far in the future (e.g. 2015 as a baseline for a subsequent 'freeze' on methyl bromide consumption by developing countries).

The setting of target dates for implementation of commitments provides similar opportunities. The ways in which a particular date aligns with relevant investment or election cycles, for example, can be expected to vary between countries. Moreover, allowing certain countries an additional 'grace period' before they have to implement can make a substantial difference to their capacity and interest in joining and complying with an agreement, as illustrated again by the treatment of developing countries in the Montreal Protocol.

Use of 'safety valves' and selective incentives

Many international agreements centred on simple and undifferentiated 'headline' commitments have been made acceptable to 'reluctant' parties by the deliberate use of exemptions, selective incentives, allowances, 'loopholes' and ambiguities. These operate as 'safety valves' or as incentives to persuade states to join agreements that they might otherwise reject as inequitable, infeasible, too costly, or contrary to other overriding concerns. They can also enable parties to side-step thorny problems and obstacles to agreement.

Selective incentives and allowances are means by which fringe benefits or burden reductions can be given to particular states or stakeholders to encourage them to join or implement an agreement. They are relatively common in substantial multilateral environmental agreements.[9] For example, LRTAP commitments only applied to European parts of the USSR. Moreover, the Soviet Union was allowed a special arrangement whereby it was committed to reduce 'transborder fluxes' instead of total emissions. Fishing quotas have often been adjusted to take into account the 'special circumstances' of some states. The Montreal Protocol allowed some transfers of national CFC production quotas among EU states to allow commercial producers in Europe to rationalize production facilities relatively cost-effectively. Similarly, the USSR was allowed 'grandfather' rights for CFC production facilities under construction until the end of 1990. Though these allowances were marginal in terms of achieving the overall aims of the ozone agreement, in practice they respectively made a major difference to the costs of implementation for European producers and played a key role in overcoming obstacles to Soviet acceptance of the ozone agreement.

Similarly, the incorporation of exemptions has been critical to the success of several major environmental agreements. The moratorium on whaling agreed by the IWC was made acceptable to Japan and some other whaling nations by the allowance for 'scientific whaling' (and also for the possibility of 'traditional whaling' by communities and indigenous peoples for whom such activities are an important part of their cultural identity). The 1972 Helsinki Convention for the Protection of the Baltic Sea

---

[9] As discussed, for example, by P. Sand, op. cit.

Environment exempted military vessels from reporting obligations and controls, to accommodate overriding security sensitivities of the Soviet Union. The exemption of consumption of ozone-depleting substances (ODS) for agreed 'essential uses' was a vital safety valve that allowed phase-outs of halons and CFCs to be agreed in 1990. Similarly, agreement to reduce consumption of methyl bromide – another ODS – was achieved only with exemptions for uses for quarantine and the pre-shipment of agricultural goods.

The Montreal Protocol's 'essential-use' mechanism is an example of a tightly defined procedure that prevents exemptions from becoming wide loopholes. However, in many cases commitments include deliberate loopholes and ambiguities in order to facilitate early agreement. Among many examples of this was the decision to avoid defining exactly what 'early notification' meant in the 1988 Convention on Early Notification of a Nuclear Accident.

Ideally, one would prefer to avoid exemptions, ambiguities and loopholes. Certainly, they should be allowed sparingly, to prevent them undermining the overall effectiveness, authority and 'defensible equity' of the basic agreement. However, experience seems to indicate that they are normally a more effective approach to addressing concerns about unacceptable or inequitable burdens or cost-effectiveness than attempting to negotiate explicit formulae for differentiated commitments. In many cases they have proved most important in overcoming the obstacles to securing an initial agreement. Once the agreement is in place, it has often proved possible to close loopholes and resolve ambiguities. In several cases, special allowances or exemptions have been designed to decline in significance over time.

Inclusion of side payments, assistance or compensation

Side payments and international arrangements to provide aid or compensation have been particularly important forms of selective incentives for environmental agreements. Most global environmental agreements have included arrangements to promote developing country access to funding, resources or technology. The same is true of regional environmental agreements involving a mixture of industrialized and developing states or of OECD states and 'countries with economies in transition'. EU environmental directives, for example, are normally formally or informally linked with side payments or resource transfers, particularly to south European states.

Similarly, joining the agreement can be linked to access to markets (or, as in the Montreal Protocol, avoidance of trade sanctions), or to access to resources. The linkage between participation in natural resource conservation and access to resources is often intrinsic to regional fishing agreements, for example.

Clearly, side payments offer an important way in which decisions about commitments to take measures to protect the environment can be decoupled from decisions about who pays for the measures. If governments are reasonably confident that equitable arrangements for payment or compensation can be arranged, they are likely to be more willing to commit themselves to relatively stringent measures. For this reason, southern EU members have been willing to go along with relatively stringent directives on the environment promoted by richer member states. Similarly, the establishment of the Multilateral Fund played a key role in persuading developing countries to join the Montreal Protocol: through the Fund, industrialized states promised to pay the incremental costs of phasing out ozone-depleting substances in developing country Parties.

However, it is important to recognize that the explicit exchange of money among nations as part of an environmental agreement has tended to be particularly difficult to negotiate.[10] It has normally been politically easier for potential donor states to negotiate trade-offs, side payments 'in kind', or simply measures involving unequal burdens. This has been the case even in circumstances where financial transfers had a good chance of achieving the desired result more cheaply. There are counter-examples, such as when the transfer would be to countries where the donor state already has a strong aid programme, or when there would be clear and direct environmental pay-offs (such as Nordic aid to Baltic states to limit sea or air pollution). Nevertheless, the point is generally true. It is sober-

---

[10] A point emphasized, for example, by Parson and Zeckhauser, op. cit., p. 103.

ing to recall, for example, the initial difficulties in persuading the United States to promise relatively modest sums to the Multilateral Fund to win developing country commitments to phase out CFCs and halons.

Exploring the possibility of package deals

The unequal costs and benefits of simple 'equal' measures in one issue area can sometimes be addressed by linking them to commitments in another issue area where the benefits are distributed differently. It is well known that linking issues or commitments together increases the scope for trade-offs and for finding package deals in which the costs of one element can be offset for each party by benefits from other elements of the package. The overall package may be more negotiable than commitments in each issue area on its own, because it has something for everyone and can pass the test of 'defensible equity'.

This approach has been used with success, but experience shows that it can have major pitfalls. The Law of the Sea negotiations provide rich lessons in this context.[10] The final treaty included a range of components in which different parties had greatly differing interests. However, it took over a decade to negotiate and longer again to come into force. Whereas the 'package deal' approach was supposed to increase flexibility, the Law of the Sea negotiations became famous for their inflexibility and unwieldiness. Agreements in one area were held up by disagreements in another. The process depended greatly on diplomatic brokers to be able to move forward at all.

Experience with this treaty and others shows that linkages between issue areas should be established only with great caution. Once different issues are 'packaged' in negotiations, it tends to be very difficult to unpackage them. Skilful sequencing of negotiating approaches is even more important in this approach than is normal. It can often be better to achieve what one can in negotiating each issue area separately before deciding whether and how to link them together.

---

[10] See, for example, J. Sebenius, 'The Law of the Sea Conference: lessons for negotiations to control global warming', in G. Sjostedt, op. cit., pp. 189–215.

Encouraging 'lead' parties to go beyond their international obligations

Another way of dealing with differences between the measures that states are prepared to take within the framework of an agreement centred on equal commitments is to find ways of encouraging some states to 'over-comply' or to unilaterally commit themselves to more ambitious targets. In this way, states which find the agreed types of measures particularly burdensome, or which are unwilling to prioritize environmental protection in this area, can be accommodated in an agreement that is only moderately stringent. Countries which are able and willing to go further would be bound by the same international obligations, but encouraged to adopt more stringent targets. In this way, differentiated measures would be promoted, without the problems of negotiating differential commitments.

It is always open to states to over-comply. Often, governments are from the outset committed to do so by domestic environmental legislation. Some international agreements go little further than to establish commitments that are close to the least common denominator of what the parties planned to do anyway. However, *effective* international agreements should be designed to promote and sustain additional measures. Thus, in this context, the challenge is to design agreements that promote over-achievement.

One way in which this has been done in international agreements has been to make it plain that additional measures are desirable, and then to establish systems that make states' environmental performance more transparent. This provides political dividends for over-achievers, raises public questions about the relatively poor performance of laggards, and can promote learning about how relevant environmental measures can be implemented effectively.

Thus, for example, the 1985 Sulphur Protocol within LRTAP called for emissions reductions of 'at least 30%' by 1994. By 1988, the LRTAP reporting system showed that twelve countries had already achieved their commitments. They announced that they would now aim for 50% reductions. This earned their governments international and domestic political credit, and embarrassed countries such as the UK and USA which had refused even to commit to 30% reductions. Similarly, after 1989, the EU, the USA and several other OECD states over-complied with their Montreal

Protocol obligations to reduce consumption of CFCs and halons, encouraging others to do likewise and facilitating the rapid strengthening of international commitments.

Governments have tended to take clearly stated political commitments seriously, even if they were not binding in international law. Thus it has been valuable for international agreements to encourage states to declare ambitious 'soft' targets. This has been a feature of environmental agreements relating to sea or river pollution, species protection and acid rain, as well as of the FCCC. However, some agreements have gone beyond this, and have helped to make commitments to 'over-achieve' more binding. For example, the 1988 NOx protocol of LRTAP allowed parties to choose from a range of baselines other than the default 1987 baseline year, provided the chosen baseline did not exceed emissions in the default year.

Agreements that allow lead states to 'bank' early environmental actions have also been helpful in this context. Not surprisingly, governments have been more willing to take a lead if they have been reasonably confident that their country can gain enduring credit for early actions if commitments are subsequently revised or strengthened. Thus, confidence that baselines were likely to remain stable in the Montreal Protocol and that future revisions of the agreement were likely to make controls more stringent provided a context in which several countries reduced their CFC and halon consumption more rapidly than they were obliged to do.

Conclusions

Experience with other environmental agreements must discourage attempts to negotiate substantial differentiated commitments under the FCCC within the Berlin Mandate, particularly if the aim is to achieve agreement in time for the third Conference of the Parties in December 1997. In many other environmental negotiations, strong arguments could be (and were) made in favour of differentiated commitments, in the interest of overall cost efficiency or to equalize burdens between parties. In practice, however, effective differentiated commitments are particularly difficult to negotiate in multilateral fora, and in any case may not be widely accepted as equitable – creating potential problems for ratification and implementation. In order to find agreements that go beyond 'least common denominator' approaches, negotiators have usually been obliged to focus on simply expressed commitments involving 'equal' measures.

The encouraging lessons from experience are that agreements based on apparently simple 'equal measure' agreements have proved more nego- tiable and effective than might be expected. One reason for this is that there has normally been a range of potentially useful types of measures available, from which negotiators can choose one which implies a rela- tively negotiable distribution of costs and benefits and yet seems uncon- trived and salient. There appears to be wide scope for this for the Climate Change Convention.

Moreover, there are many ways in which an agreement based on simply expressed, undifferentiated 'headline' commitments can be designed, adjusted and qualified to accommodate the differing interests and concerns of a wide range of parties. All or most of the ways for doing this that have been outlined in the previous section appear to be potentially available to negotiators in the Ad Hoc Group on the Berlin Mandate (AGBM).

On the basis of present scientific evidence, negotiations within the AGBM must be seen as only an early step in a long-term process. Commitments negotiated in time for COP 3 will need to be revised and strengthened in subsequent years. Whereas the lessons from other agree- ments indicate that there is little prospect of negotiating effective differen- tiated commitments by December 1997, in the longer term the prospects may be brighter.

Experience from the agreements to combat acid rain may provide some of the most relevant lessons on ways of developing differentiated commit- ments over time. As discussed, the first substantial commitments were based on requirements for equal percentage cuts in emissions (but making use of several of the approaches outlined in the previous section for mak- ing them more negotiable and effective). Between 1979 and 1994, however, participants in LRTAP became familiar with the demands of implementing the regime, one another's concerns and constraints, and the acid rain mod- els (particularly the RAINS model). This made it possible to switch to the 'critical loads' approach in 1994, with its implications of differentiated commitments and greater cost-efficiency.

In spite of the substantial differences between the climate change and acid rain issues and agreements, I think that this example provides important lessons for ways to develop FCCC commitments effectively. It indicates that it should be more possible to agree complex differentiated commitments in the FCCC as the regime becomes well established, detailed common understandings have developed of the problem and of responses, and simple, equal commitments have both been agreed and begun to be effectively implemented. If the negotiations under the Berlin Mandate are successful, the prospects for negotiating more complex and differentiated commitments after 1997 could improve, with potentially substantial benefits. The process of developing the frameworks, institutions and shared understanding necessary to make this possible can be promoted immediately.

Chapter 4

On the differentiation of quantified emission limitation and reduction objectives (QELROs) for Annex I countries

Michael Grubb

General considerations regarding differentiation

Introduction

This chapter discusses some issues surrounding the thorny and highly political question of whether 'quantified emission limitation and reduction objectives' (QELROs) by Annex I Parties could or should plausibly be differentiated, and if so, how. It is not my purpose to suggest any specific numerical targets, or to enter into the broader debate about the value of emission targets *per se*, but this section offers some short general comments, before turning to the central questions, which are defined primarily by issues surrounding energy-related CO_2 emissions.

Some commentators have questioned the appropriateness of quantified emission objectives in any form. However, the current 'aim' in the Convention has in many countries played an important role in providing a focus for national efforts and an assessment of progress. An important limitation has been its short timeframe; while political systems have a fairly short time focus, energy systems, in particular, have long ones. The aim and timescales (from 2005 to 2020) of the Berlin Mandate negotiations on QELROs thus seem quite appropriate.

There is a substantive debate on the form that any emission targets (to use the term broadly) should take. This chapter does not seek to add to what has already been said about the appropriate firmness of language in which emission targets might be phrased – where on the spectrum from hard, legally enforceable targets to soft, indicative targets may be best.[1] I

---

[1] For an overview of this debate see the Rapporteur's report of discussions at an earlier RIIA workshop, in M. Grubb and D. Anderson (eds), *The Emerging International Regime for Climate Change: Structures and Options after Berlin*, RIIA, London, 1995, and references therein.

will simply assume that the targets are intended to be 'hard enough' for questions of differentiation – and related questions of whether countries may reasonably meet the targets – to be relevant.

Another issue to be considered is the coverage of greenhouse gases: should QELROs for Annex I focus on the main Annex I greenhouse gas (CO_2 from fossil fuels), on several different gases individually, or on a 'basket' of different gases? This raises questions of comparability, monitorability, control and other factors. I am not aware of any study that seriously examines the trade-offs involved, but if there are realizable benefits to be made by expanding the net of allowable emissions, one approach could be to establish a list of quantifiable and comparable sources and sinks, along the lines previously suggested in an article by David Victor, Chris Hope and myself.[2]

The paradox of differentiation

Since the ease of achieving any given emission target is likely to vary considerably across countries, there are two big drawbacks to negotiating an agreement that requires all countries to address the same target:

- It is not efficient, because it involves widely varying marginal costs of control.
- It is likely to be environmentally relatively ineffective; negotiations will tend to bring the target down towards the 'lowest common denominator' because that is the only number that will gain full adherence to the agreement, which may leave some participants having to do relatively little.

Despite this, many international environmental agreements have in fact focused upon simple, flat target agreements, at least for industrialized countries (as with the Montreal Protocol). This is for several reasons.[3] It is much

---

[2] Michael Grubb, David Victor and Chris Hope, 'Pragmatics in the Greenhouse', *Nature,* 354, pp. 348–50, December 1991.

[3] For discussion see Edward A. Parson and Richard J. Zeckhauser, 'Equal measures or fair burdens: negotiating environmental treaties in an unequal world', in Henry Lee (ed.), *Shaping National Responses to Climate Change: A Post-Rio Guide*, Island Press, Washington DC, 1995, pp. 81–114.

simpler, it gives an innate (if often inaccurate) sense of equity, and it is more likely to achieve agreement than the treacherous waters of differentiated negotiations.[4] In the case of climate change, raising the question of differentiation among countries potentially opens a Pandora's box of special pleading. The economic systems that emit greenhouse gases are extremely complex. Each country has a different evolution, different resource endowments, different current situation, and different future expectations and ambitions, that could affect greenhouse gas emissions. If differentiation is taken to mean trying to negotiate different numerical targets for each individual country according to these circumstances, the scope for never-ending distributional arguments could sink the negotiations.

In some ways, focusing on one flat target for all also gives a better incentive structure. Negotiations that are differentiated country by country give a direct incentive for each country to exaggerate the difficulty of achieving any given target, whereas in flat target negotiations many countries can afford to be more optimistic because they know the actual result is likely to be less demanding. Indeed, they have some incentive to be more optimistic about what is possible because any tightening of control will apply to all, and hence achieve greater environmental control.

For these reasons, any steps towards differentiation need to be taken with great caution. Nevertheless, given the breadth of differences within Annex I, it does appear that any feasible undifferentiated agreement, based on national 1990 emissions, is likely to be relatively inefficient and ineffective. The common goal might have to be largely determined by what Australia, Canada and Norway consider to be achievable, for example, if they are to be party to it. Hence the search for feasible approaches to differentiation.

---

[4] But see Owen Greene, in Chapter 3 of this volume, for an argument that flat-rate reductions can be seen as equitable, and also for a more detailed discussion of how flat target agreements may in practice accommodate some differentiation, and how they have in some cases evolved later into more complex differentiated agreements.

The relationship between differentiation and emissions trading

One approach to the problem of differentiation is to place the emphasis on the potential for emissions trading; countries could, in one way or another, be free to alter their initial emissions 'allocation' through bilateral trade with other countries. In principle, this could be implemented in many different ways, at intergovernmental or industry level.

This reduces the problems associated with countries being bound by a flat emissions target. It enables a more efficient outcome, because countries facing higher abatement costs may purchase 'permits' from those for which reductions below the flat target are much cheaper. By the same token, it makes agreement on a flat target politically more feasible, by reducing the costs for those countries and generally giving greater flexibility in implementation of commitments, which is very useful given the inherent uncertainties in energy and emission projections. Allowing 'banking', or other ways of trading over time, may further increase the flexibility. One objection sometimes raised is that such a system, once created, could not evolve with new targets or expand to include new entrants. These, along with many 'structural' objections, do not appear to be valid.[5]

But although trading may ease the problems associated with flat (undifferentiated) emission targets, it does not resolve the overall problem of differentiation. Instead of focusing the debate upon allocation of emission targets, it focuses it upon the allocation of tradeable 'property rights' – the right to emit CO_2 – which may then prove very valuable. The fundamental question of what constitutes a fair and feasible allocation remains. Trading might just possibly allow an undifferentiated (flat) target to be reached across Annex I countries in the next phase of negotiations – though some countries would raise vociferous objections, and questions over the role of countries with economies in transition would remain – but it would not in any way help to address the fundamental and long-term dilemma of differentiation.

---

[5] See *Climate Change: Designing a Tradable Permit System*, OECD, Paris, 1992.

Indexing emissions

One solution proposed for the paradox of differentiation is to seek *indices* of reasonable emissions, or of changes in emissions from a base year. Various suggestions have been made: that emissions could be related to GNP or GNP growth, population or population growth, land area, or average temperature as a driver of energy consumption, for example. Other contributions to the workshop have offered and considered some examples.[6]

Some of these appear more credible than others. Land area, for example, is a double-edged influence: how should one weigh the potentially greater transport requirements against the greater opportunities for renewable energy, for example? So is temperature: how does one compare a temperate coastal climate with a deep continental interior that may have the same annual average but face massive fluctuations between freezing winters (needing heating) and baking summers (requiring air-conditioning)?

GNP and population have some claim to be more fundamental, but they still face the problem that indexing is crude: it is seeking a common driver for emissions that derive from energy systems which are complex and diverse within countries as well as between countries. However, there may be possibilities that may prove particularly interesting in conjunction with other options discussed below.

Overall, however, indexing alone seems unlikely to solve the differentiation problem – fundamentally, in fact, because it is not really an instrument of differentiation, in the sense of addressing really fundamental differences in economic structure or patterns of development.

Principles of group differentiation

If differences exist that render a common target, as a metric of effort or responsibilities, demonstrably ineffective or inequitable, it may be necessary to consider explicitly recognizing this by dividing Parties into more than one group in terms of the nature of their commitments. The divide between Annex I and non-Annex I in the Convention is an example of such an obvious, essential and unquestioned divide.

---

[6] See in particular Kawashima, Chapter 5 of this volume, for a detailed account of such a means of differentiating commitments.

The obvious principle on which such differentiation should be based is that only the biggest differences should be captured, into as few groups as possible, and that these groups are defined in ways that are based on identifiable cleavages. Membership of different groups would ultimately have to be voluntary; countries would decide for themselves to which group they were best suited, taking into account international responsibilities. The key is to find a minimum number and definition of groups to allow for tolerable efficiency, equity, and effectiveness in the overall effort.

Major cleavages among industrialized countries

The Climate Convention, and most discussion of future QELROs, tend to be based on an idea of changes relative to a 1990 base year. For the developed OECD economies, there are clear reasons for this that I need not elaborate here, beyond the fact that their essential, common characteristic is continuity in a mature economic structure before and since the base year. Within Annex I, however, there are two other broad kinds of economic situation.

The most obvious concerns the 'economies in transition' (EIT) countries of central and eastern Europe. These countries have undergone radical transition, associated with economic contraction and steeply declining levels of emissions. Moreover, they were at various stages of this process in 1990. This inherently makes 1990 a base year of questionable relevance.

There is, however, another category, which consists of countries that in 1990 were (and that still are) undergoing rapid industrialization. Within Annex I, this includes several south European countries, and it has posed no overt problems for differentiation under the Convention, if only because these countries are covered under the EU umbrella commitment. Yet within the OECD, such countries now also encompass Mexico and soon will include South Korea. These two countries have stayed out of Annex I in part because a commitment which takes 1990 as its base year makes it almost impossible for them to consider joining it.

Yet it may be politically very valuable to find a categorization that makes it possible for these countries to join equitably in emission commitments that are somewhat more specific than those for developing countries in general – perhaps involving QELROs defined on a different basis that allows for appropriate, but not wholly unlimited, emissions growth.

Does this imply that a new agreement on QELROs should contain three principal categories of countries? Perhaps. Just possibly, however, it may be feasible to achieve the main objectives with only two categories, because of the nature of the EIT countries' transformation.

Some EIT countries have argued that their emission commitments should be related to a base year prior to 1990, on the grounds that for some, emissions started declining rapidly from a peak in 1987 or 1988. The fundamental problem is that such back-dating, while politically expedient, is not rooted in any economic basis that has a connection with the present. The real desire of EIT countries is not to return to their former economic structures; it is, rather, to gain space to expand their new, market-based economic systems. For this they believe their emissions must grow, perhaps substantially, from the level at which their economies are 'bottoming out' during the 1990s. And that is the point of economic continuity upon which a QELRO's base year, allowing for such growth, could more rationally be defined – if indeed it is to be related to a base year at all.

And that, at least qualitatively, is not so different from the situation faced by the newly industrialized countries (NICs). The quantitative differences may simply be too large to make it possible to group these two categories together – this would require empirical analysis of whether the economic structures of newly industrialized countries and the transition economies are similar enough to make any such grouping feasible. Such analysis might also point to an entirely different basis for QELRO commitments – for example, a long-term per capita commitment. Even if the NICs were unwilling to consider joining such a regime at this stage, it could still point towards seeking a regime for EIT countries that could later encompass NICs. The suggestions are highly tentative; but such issues at least seem worth exploring.

Sectoral differentiation

The other kind of differentiation, which has received little consideration in the debate so far, is sectoral. Greenhouse gases come from many different activities. These different sectors vary enormously in their growth patterns, ease of control, institutional characteristics, and diversity or similarity between countries.

Any solution to the differentiation problem must ultimately be based on an understanding of these different sectoral characteristics and driving forces. The situation of, and concerns raised by, different countries largely reflect differences in their economic structure. Any credible approach to differentiation must ultimately reflect these underlying factors.

Three broad approaches seem available.

- *Sectoral analysis* could be used simply as a basis for estimating differentiated total national obligations – as an input to calculating national totals, and no more. This relates directly to the broader discussion of the economic basis of differentiation, with all its claims and counterclaims, pros and cons.
- *Sectoral agreements.* It has been suggested that agreements should be reached governing the total obligations on different sectors internationally, e.g. on emissions from electricity production, chemicals, etc. This seems a non-starter for institutional reasons: there is no legal entity that can negotiate or be held responsible for such agreements.
- *Disaggregated national commitments.* National obligations could be defined for two or more sectors separately. The underlying issue here is similar to that faced in geographical differentiation/grouping proposals: it could be overdone, leading to a proliferation of unmanageable negotiations. Trying to split national obligations into too many different, interrelated sectors would ultimately be inefficient, ineffective and self-defeating. But there are valid questions of whether there is some degree of separation, whether there are major cleavage lines into different economic sectors that can be treated with some independence from one another.

Sweden has already proposed that agricultural emissions should be considered separately from industrial emissions.[7] This makes sense, and so might be one further step. The core negotiating problem is that energy overall is just too big and too diverse, and there is at least one natural cleavage that may simplify the negotiating problem. That is the separation between energy used for economic production and that used for private consumption.

Industrial and service emissions compared with domestic and transport emissions: distinguishing features

Focusing upon energy, the distinction is in most cases quite clear. Emissions from production are those that arise primarily from manufacturing and energy transformation activities: steel, chemicals, etc., together with electricity production, refineries operation, gas pumping, etc. Emissions from economic consumption arise from energy use in domestic dwellings, private transport, etc. There are a few grey areas, notably in freight transport and parts of the service sector. These definitional issues are addressed in one of the most comprehensive studies of energy in development that spans analysis of developed, developing and middle-income economies.[8] This study separates production from consumption and argues that this is essential to a real understanding of the role of energy in economic development.

For clarity, I consider the division to be between *industrial and service* emissions, on the one hand (including energy production and transformation industries such as power generation), and *buildings and transport* emissions, on the other; a shorthand can be *industrial versus consumption-related* emissions. I assume emissions to be allocated where they occur: all emissions from power generation, for example, are industrial emissions, wherever the electricity is consumed. Trying to shift responsibilities down-

---

[7] Swedish Commission on Climate Change, *The Global Climate Change*, Swedish Environmental Protection Agency, 1996.

[8] 'Energy planning should clearly distinguish energy demand in the production and consumption sectors as they need completely different actions for energy demand management and energy supply' (G. Leach et al., *Energy and Growth*, Butterworths, 1986, p. 164).

stream (e.g. making electricity consumers responsible for emissions from power generation) or upstream (e.g. making oil companies responsible for emissions that occur when petroleum is burned in cars) in my view leads to a logistical and institutional morass for policy-making, and should be rejected. In the OECD, industrial emissions in this sense account for slightly more than half the total CO_2 emissions.

Key differences between production-related and consumption-related energy demand are summarized in Figure 4.1, which also illustrates graphically the different trends in production-related and consumption-related energy demand over the 1970s (in 1971, 1976 and 1981) for the five developed countries analysed in the *Energy and Growth* study. In total there seem to be at least six important distinctions between the two.

(a) Driving forces and indexing Separating industrial from consumption-related emissions may help with 'indexing'. Manufacturing activities tend to be quite closely related to GDP. Consumption-related emissions can be expected to be driven principally by population and by personal income and consumption habits; they would need to be indexed differently, with reference to population and/or more sophisticated indicators such as building area and distance travelled. Obviously the two are related, but separating industrial production from consumption-related emissions may enable more directly relevant (and more equitable and efficient) indexing, if this is to be considered at all.

These differences suggest that there may be several potential benefits to a process in which negotiations address production-related and consumption-related emissions separately, establishing QELROs separately for each. Though politically such negotiations would of course be linked, the following considerations suggest that such separation may make it easier to reach an effective and efficient agreement.

(b) Differing trends As suggested in Figure 4.1, industrial energy consumption has followed quite similar trends in developed countries (also in most middle-income countries), even during the period of massive upheaval of energy prices in the 1970s. As shown in studies by Schipper et al. that extend such analysis through the 1980s, in OECD countries

Figure 4.1: Economic production versus consumption trends

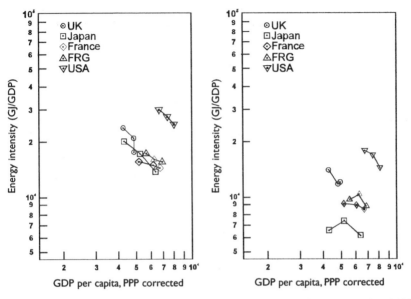

Note: The three points for each country indicate data successively for 1971, 1976 and 1981.

Industry, conversion and services	Personal transport and domestic
Parallel trends in OECD	Diverse trends in OECD
Internationalized technology	Domestic culture and behaviour
Driven by GDP, manufacturing structure and resource endowments	Driven by consumer habits, consumer technologies and infrastructure (often public)
Policy concerns about international competitiveness	Policy concerns about electoral impact, political feasibility
Mostly large corporate actors	Individuals
Most investments, 20–50 year lifetime	Investments range 1–100 year lifetime
Emissions principally from coal and gas	Emissions principally from oil products and gas

Source: G. Leach, L. Jarass, G. Obermair and L. Hoffmann, *Energy and Growth: A Comparison of 13 Industrial and Developing Countries*, Butterworths, Guildford, 1986.

energy intensity for manufacturing has declined at similar rates.[9] This is not surprising; it reflects the international nature of manufacturing technologies and the relative rapidity of transfer of related technologies throughout the developed world.

Energy for consumption, by contrast, has gone in different directions in different countries, and reversed the direction of trends in some. It is also at a less 'mature' phase of development; whereas energy intensity for production has declined steadily in all OECD countries for several decades, the stage of development of consumption-related demand varies widely (compare, for example, Japan with the USA). Thus it would not be surprising if a different, and potentially more complex, regime needed to be developed to address consumption-related emissions equitably.

The studies by Schipper et al. have extended the analysis of such trends for manufacturing to more countries and explicitly to CO_2 emissions, with results shown in Figure 4.2. This illustrates that even for CO_2, there have been quite similar trends in eight out of the ten countries considered. In the two outliers, the cause is very clear: Sweden's development of non-fossil electricity sources is contrasted with Denmark's development of coal power, both in response to the oil shocks. Accordingly, an index based on carbon emissions per unit GNP for manufacturing emissions would have rewarded the former and penalized the latter without penalizing the economic growth itself. This would be entirely appropriate for tackling a carbon-related problem; and Denmark's subsequent change of policy – towards promoting renewable electricity sources – would be correspondingly rewarded as the effects work through and bring its carbon intensity down.

(c) International competitiveness versus electoral impact Another issue flows from this difference. The difficulties that governments have concerning the negotiation and implementation of emission restrictions stem from two main sources. One is fear about the impact on the international competitiveness of their industry. The other concerns electoral popularity.

---

[9] L.J. Schipper, R. Haas and C. Sheinbaum, 'Recent trends in residential energy use in OECD countries and their impact on carbon dioxide emissions', *Journal of Mitigation and Adaptation to Global Change* (forthcoming, 1996).

Figure 4.2: OECD-10 manufacturing carbon intensity

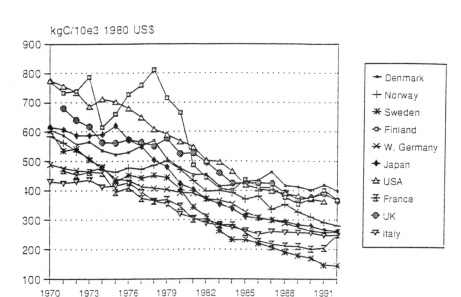

Source: Schipper et al. (see note 9).

These reflect, respectively, issues of production and consumption. Separating the two could therefore have big political advantages. Negotiations about production-related emissions could focus squarely on the issue of maintaining a 'level playing field' in terms of impact on industrial production costs. Negotiation of consumption-related commitments need not be entangled in such issues, and could focus on comparison of consumption activities and the opportunities for abatement in ways that are electorally feasible given the culture of the countries concerned.

(d) Institutional structures and implementation Differences extend also to the institutional structures concerned. Production-related emissions are generated almost entirely through the activities of corporate actors. Consumption-related emissions are the product of individual consumption choices. Consequently, the implementation issues differ radically and, accordingly, entirely different policy instruments may be appropriate and feasible. Major contributions to the control of production-related emissions

may be made through mechanisms such as effective corporate voluntary agreements, tradeable emission permits, or other instruments that require the participation of large institutional actors. For consumption-related emissions such instruments may be infeasible, and education, efficiency standards or tax incentives may be far more appropriate and effective.

Emission commitments for production could themselves be internationalized. For example, if some countries implement production-related emission constraints using tradeable permit schemes, an emissions agreement could be structured to allow for this flexibility for international exchange of permits among these countries. For various reasons, countries may be less willing or less able to 'internationalize' consumption-related emission constraints at this stage, partly because of the institutional differences and political sensitivities involved.

Similar comments apply to the timescales involved. The timescales associated with industrial investment are somewhat more homogeneous than those associated with domestic and transport emissions. Recent debates over the timing of constraints have demonstrated that this too is an important consideration.

(e) Fuel mix Furthermore, the mix of fuels between these sectors varies considerably. Industrial emissions are dominated by coal and gas, with some fuel oil. Buildings and transport emissions are associated with refined oil products and gas. Delineating negotiations between the two may thus also simplify the fuel-industry politics of the negotiations.

(f) Relationship to geographical differentiation Finally, separating industry-related from consumption-related emissions may contribute to easing the problem of geographic differentiation. This can best be illustrated with reference to specific examples. For instance, part of Norway's difficulty is that it starts from a very low 1990 base of emissions intensity for production, because its electricity system is 100% hydro power, but there is great pressure for increasing emissions from the rapid growth in its offshore industries (as well as in transport). Unlike some other OECD countries, Norway has no opportunity to offset these pressures against reductions from electricity sector emissions. A target that specified rates of

ment in carbon intensity from production, combined with separate commitments for consumption-related emissions, could accommodate these realities far better than a flat national emissions target, without necessarily sacrificing (indeed, potentially improving) overall control.

Possible features of sector-differentiated QELROs: a summary

The discussion points to different possible features of an agreement covering industrial and service emissions as compared to one covering domestic and transport emissions (agriculture/land use would presumably have different features again). These may be summarized as follows.

Industry and services

- Possibility of indexing to GDP growth (e.g. X%/yr decline in carbon to GDP ratio).
- Strong international orientation: possibly design with international flexibility.
- Related to industrial investment timescales; frame with reference to industrial concerns (especially electricity and industries with heavy energy use).
- Actors capable of long-term planning and evaluation; possibility of designing some time-flexibility (e.g. cumulative over defined period) if credible enforcement is possible.
- Tradeable emission permits could be a natural tool of implementation, at least for most sub-sectors and the great majority of emissions.

Domestic and transport

- Possibility of indexing to population/population growth, or more targeted indicators such as floor-area times degree-days,[10] for buildings, and passenger kilometres for private transport.
- Less point in international or inter-temporal flexibility; frame with reference to cultural/electoral sensitivities.

---

[10] Degree-days are an internationally recognized tool of energy analysis, to express the extent to which temperature in a given year differs from a reference level, as an indicator of potential heating needs.

- Soft law framework initially for QELROs?
- Focus more on policies and measures?

The suggestion that QELRO negotiations should actually be delineated in this way is advanced tentatively. Separating energy-related industrial and service emissons from consumption-related emissions (with agriculture/land use as a third category) would remove the neat simplicity of a flat national emissions target, and introduce some definitional complexities. It would also represent a loss of efficiency compared with the theoretical ideal (though this is somewhat academic if effective and efficient country-by-country differentiation is in fact impossible for the reasons sketched above). But the six potential benefits outlined here do seem sufficient to justify further exploration of the pros and cons of taking such an approach in the context of QELRO negotiations.

Chapter 5
The possibility of differentiating targets: indices and indexing proposals for equity

*Yasuko Kawashima**

Setting emissions targets

There have been debates among Annex I Parties concerning the kind of QELROs which should be adopted for the period beyond the year 2000. Should they be uniform or differentiated? If differentiated ones are preferred, then what are the criteria by which they should be differentiated? Although these discussions are indispensable to frame a protocol or other legal instrument to the Framework Convention on Climate Change, few studies have actually calculated the allowable CO_2 emissions under each target-setting proposal. This chapter focuses on the possibility of differentiating QELROs by indexing them to major national circumstances. Effectiveness, equity and efficiency are three important principles for such differentiation. Taking the outcome of this calculation into account, as well as the limited time available until the adoption of a protocol (COP 3, in 1997), I propose a negotiation procedure for QELROs.

This chapter deals only with various ways to differentiate QELROs, but there are other important elements in the expected protocol, such as commitments on policies and measures.[1]

* Appreciation is expressed to the Japan Environment Agency, especially to Hironori Hamanaka, Hikaru Kobayashi and Katsunori Suzuki, as well as to the many scientists who offered critical suggestions – most of all to Naoki Matsuo of the Institute of Energy Economics and Tsuneyuki Morita of the National Institute for Environmental Studies.
[1] For an interesting proposal concerning such commitments, see Japan Centre of International and Comparative Environmental Law (NGO), *Possible Protocols on the FCCC*, Tokyo, 1996.

Effectiveness, equity and efficiency

The volume of allowable emissions for each country may be determined by three principles: effectiveness, equity and efficiency. This section defines what is meant here by these principles.

Effectiveness is defined as setting a limit on the total amount of emissions at a level which will prevent dangerous anthropogenic interference with the climate system. This limit should basically be provided by scientific findings concerning climate change, rather than by expected future demand for fossil fuel consumption. It does not necessarily mean that QELROs can only be discussed if there is no scientific uncertainty. There is adequate information available to set such a limit. The 'emissions corridor' which was presented at an informal workshop on QELROs is one idea to set a limit to world emissions.[2] It is important that this emissions path is determined, so that the protocol will be compatible with the ultimate objective of the FCCC. In this sense, QELROs such as 'stabilization or reduction of emissions per GDP' are not suitable, because the commitment can be fulfilled while emissions keep on growing as long as GDP growth is greater than emissions growth. This paper assumes that the total emissions limit is given, and focuses only on the means to distribute that figure for total emissions among countries.[3]

After this total emissions limit is set, the next step is to determine an acceptable burden-sharing rule. In the long run, the amount of emissions may be allocated under an equity principle. However, what is the most equitable, or fairest distribution? This is the focus of this chapter and will be discussed in the next section.

Even if countries can agree on an equity principle, the distribution of emissions under this principle may be possible only as a long-term aim because we cannot neglect the present emissions levels. Electricity generation plants, railroads, district heating systems, etc. are critical elements

---

[2] J. Alcamo and Eric Kreileman, *The Global Climate System: Near Term Action for Long Term Protection*, National Institute of Public Health and the Environment (RIVM), report no. 481508001, 1996.

[3] An interesting discussion of other kinds of targets is introduced in N. Matsuo, *A Proposal for the Protocol(s) on Climate Change*, Presentation at the Workshop on QELROs, Ad Hoc Group on the Berlin Mandate meeting, Geneva, 28 February 1996.

for the determination of CO_2 emissions levels, but it would involve significant cost if they were to be replaced in a short timeframe. It is not efficient, or cost-effective, to cut emissions by a large percentage in a short period of time. We have to take into account the status quo and determine the emissions path that would allow all countries to achieve gradually the ultimate distribution of emissions under equity.

There are three main approaches to sharing emissions reduction among countries. The first is to accept current emissions levels as they are and have all countries reduce their emissions at the same rate. This is what is now dubbed 'uniform reductions'. The major disadvantage of this approach is that it is not an allocation based on equity. However, at the same time, this may also be considered an advantage because it leaves out debates over equity when the time available for negotiation is limited. Therefore, it is one viable option.

The second approach is to allocate emissions in proportion to an equity-distribution ratio. In this case, the expected equity allocation will be achieved instantly, but some countries may have to reduce their emissions drastically while others may increase their emissions. This is an inefficient approach, which does not consider current emissions.

The third approach is to take the difference between the current and the ultimate distribution of emissions (based on a concept of minimum reasonable need, on which more below), and allocate emissions reductions in proportion to that difference. This is burden-sharing in accordance with the amount by which emissions exceed the minimum need. I call this approach 'reduction of excess emissions objectives' (REEOs). With REEOs, countries whose current emissions exceed their minimum needs more than other countries would subsequently have to reduce their emissions by a greater amount. In most cases, countries with relatively high excess emissions are those that have had large emissions in the past, so this fulfils the 'polluter pays' principle (PPP) as well. At the same time, reductions would be made from current emissions, so the reduction schedule would be more gradual than the previous option. The following discussion uses the REEO to estimate emissions targets for each country.

Another proposal similar to REEO is to start at current emissions levels, set a long-term per capita target, and make a gradual linear reduction in emissions towards this long-term target.[4] This is a worthwhile proposal to consider, but it would be difficult to change either the ultimate long-term target, once fixed, or the allowable emissions level of each country when new scientific findings are obtained or if population growth did not follow the prediction. It is better to leave the flexibility of QELROs so that they can reflect the latest findings on climate change.

Equity criteria and distribution of emissions

This section deals with the second issue in the previous section – equity. First, four possible equity criteria are introduced to determine a country's 'minimum reasonable need' for emissions. Total emissions are then distributed acording to each of those criteria.

Equity criteria

Per capita emissions This is often regarded as the primary equity rule. Per capita emissions vary considerably even among developed countries. There may be a variety of reasons for this, but we could say that the emissions level of the country with the lowest per capita emissions is the minimum reasonable need.

Per GDP emissions Countries want to achieve emissions reduction while still maintaining sound economic growth. In order to decouple CO_2 emissions growth successfully from economic activity, per GDP emissions is an appropriate index to measure the energy intensity of the economy. Again, CO_2 emissions per GDP vary considerably among OECD countries. Under this criterion, the minimum reasonable need is the emissions level of the country with the lowest per GDP emissions.

Per capita energy consumption (carbon intensity) Countries differ in their sources of primary energy consumption. Some countries use fossil

---

[4] Swedish Commission on Climate Change, *The Global Climate Change*, Swedish Environmental Protection Agency, 1996.

fuel, especially coal, for electricity generation, whereas in other countries most electricity is generated by hydro or nuclear power plants. Even if two countries enjoyed the same level of energy consumption, one country may have lower CO_2 emissions than the other if their primary sources of energy are different. To allow consideration for countries with less opportunity for renewable energy, the minimum reasonable need is defined as the energy consumption level of the country with the lowest per capita energy consumption.

Multi-criteria This calculation rule is developed to take into account, as far as possible, major national circumstances that could be appropriate reasons for differentiating QELROs. The above three principles are integrated together with two or more circumstances into one index to determine the minimum reasonable emissions of each country:

- *Population:* a country with larger population will be allowed greater emissions.
- *GDP:* a country with higher economic activity will be allowed higher levels of emissions.
- *Carbon intensity:* a country with less opportunity for renewable (hydro, geothermal, etc.) or nuclear energy will be allowed more emissions.
- *Temperature:* a country in a colder region will be allowed more emissions from the residential sector for heating. There is a clear relationship between energy consumption per capita in the residential sector and the temperature of the coldest month. Therefore, countries that suffer cold winters will be allowed additional emissions.
- *Area:* a country with a larger land area should be allowed more emissions for transportation. Although figures suggest that it is difficult to establish a clear relationship between land area and transportation energy, the three largest OECD countries (Australia, Canada and the USA) use considerably more energy for transportation than do other industrialized countries.

The calculation of minimum reasonable needs therefore has to disaggregate emissions by sector. The above five elements are considered only in sectors where they are regarded as relevant (Table 5.1). Size of population is considered only in residential and transportation sectors because emissions from other sectors are more closely related to economic activities. Carbon intensity is considered in all sectors, which means that countries that generate electricity from coal are differentiated from countries that do so from nuclear or renewable energy. Temperature is considered only in the residential sector because it is related to emissions from heating buildings. It is difficult to define temperature since climate differs even within one country. Here, the average temperature of the coldest month in the capital city of each country is selected as the index. The size of a country is considered only for emissions from the transportation sector. Since the amount of electricity generation depends on the final consumption of electricity, CO_2 emissions from the energy conversion sector are distributed to the other five sectors mentioned above, according to final consumption of electricity.

Table 5.1: Factors which are considered in the respective sectors

	Industry	Agriculture	Residential	Commercial	Transportation
Population	—	—	√	—	√
GDP	√	√	—	√	—
Carbon intensity	√	√	√	√	√
Temperature	—	—	√	—	—
Area	—	—	—	—	√

√ = considered
Source: Author.

Emissions distribution calculated from equity rules

Although the Berlin Mandate calls for the participation of all Annex I Parties, this chapter gives calculations only for major emitters of the OECD country Parties, for which data are available. To facilitate comparison among different proposals, all targets are set so that total emissions from all the OECD country Parties would be 10% less than in 1990. Target year (2005, 2010 or 2015) is another critical element but is not considered here, on the assumption that it will be determined basically by scientific findings as the total emissions limit is fixed. Figures 5.1–5.4 show the emissions reductions which would be required by each country under each equity rule.

Comparison of burden-sharing rules

The following findings are obtained by comparing the distribution of emissions under four different equity criteria:

- For countries such as Finland, France and Sweden which use a substantial amount of renewable energy and nuclear power, differentiation that considers carbon intensity rather than CO_2 emissions itself is disadvantageous (here I use this term to mean that countries have to reduce their emissions by a greater amount relative to other criteria). On the other hand, Australia benefits from rules that consider energy consumption as their differentiating basis.
- For Denmark, Germany, Italy, Japan and Spain, any kind of differentiation is advantageous by comparison with uniform targets. On the other hand, Canada and the USA will have to reduce more under differentiated targets than under uniform targets, and if they prefer the former, they would need to find other criteria for differentiation.
- Targets set under a single criterion are beneficial to the Netherlands, New Zealand and the UK, but multi-criteria rules become disadvantageous for those countries, mainly because they are neither large nor very cold in winter.
- Distributions under multi-criteria that consider five different factors are more similar to uniform targets than those under single criteria. The more national circumstances are considered, the closer QELROs get to uniform targets.

Figure 5.1: REEO from the lowest per capita emission level

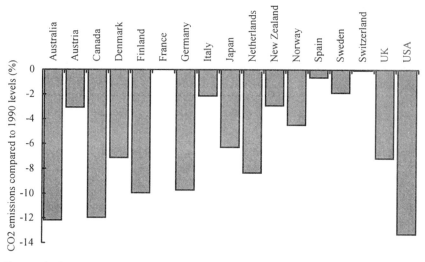

Source: Author.

Figure 5.2: REEO from the lowest per GDP emission level

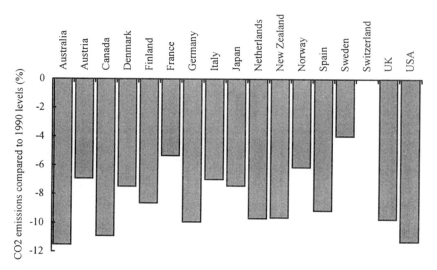

Source: Author.

Figure 5.3: REEO from the lowest per capita energy consumption

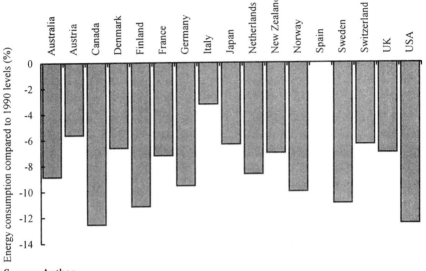

Source: Author.

Figure 5.4: REEO from multi-criteria, minimum need

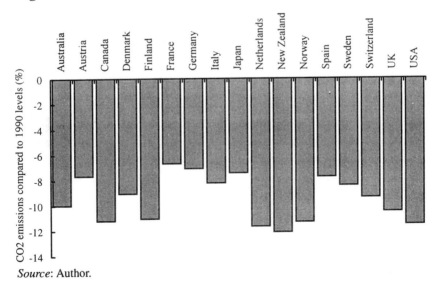

Source: Author.

There are reasons why countries have difficulties in reducing their CO_2 emissions by more than other countries, and there are ways to take these difficulties into consideration. However, the particular countries that will face higher rates of emissions reduction than others differ according to which equity criteria are chosen. In the real negotiations, therefore, what is likely to happen is that even if countries agree that differentiation is important, it would be difficult to agree on the criteria for differentiation. With the limited time available for negotiation, I would like to propose the following negotiation process:

(1) All Annex I countries clarify their positions regarding their preference for differentiated or uniform QELROs.
(2) Countries that support differentiated QELROs offer one or two ways of differentiation as their agreed proposals.
(3) If (2) succeeds, then countries that support uniform targets will examine whether it is possible for them to accept such differentiation. If countries supporting differentiation cannot agree on how to differentiate, then all Annex I countries will be committed to uniform targets.

I do not intend to conclude by recommending a certain equity criterion as the 'best solution'. Even if there is such a solution as 'the most equitable criterion', it is not likely to be the most acceptable criterion to all countries. Since the necessary suggestion for the Ad Hoc Group on the Berlin Mandate (AGBM) process may be an 'acceptable solution', this chapter has put forward possible approaches for differentiation, and has proposed a negotiation procedure for QELROs, rather than a particular criterion to be used. This procedure will prevent the final decision from being delayed by negotiations merely on the differentiation of QELROs. The time is limited, and it is important to develop climate policies without further delay.

Chapter 6

Options for EIT Annex I Parties and perspectives on greenhouse gas emissions reduction

Maciej Sadowski [*]

The differentiation of commitments among Annex I Parties is one of the most controversial issues in the process of negotiating the Berlin Mandate. Several possible criteria have been presented by the Parties. One proposal is to divide Annex I Parties into two groups: OECD countries and Economies in Transition (EIT) countries. Another is to establish a separate protocol for this second group. However, for some EIT countries such an approach creates substantial difficulties, because the definition of an EIT covers countries with different levels of greenhouse gas (GHG) emissions, different social problems and different potentials for future development.

An example of such differences is shown in Figure 6.1 (changes in GDP from 1990 until 1993). Two groups of countries may be distinguished: the Czech Republic, Hungary, Poland and Romania, which had increasing GDP during this period, and the other EIT countries, where GDP was still declining.

All EIT countries have, however, one common denominator – a reduction in CO_2 emissions from energy-related sectors (Figure 6.2), connected with economic changes after the collapse of the former political system. In fact, the reduction by EIT Parties during the period 1990–93 resulted in a general decrease in the emissions of Annex I Parties by 5.2% (Table 6.1).

The reasons for this decline vary. For the first group of countries the decrease in CO_2 emissions began as result of the recession, but now it is related to the restructuring of their economies. For the second group of countries the reduction in emissions is probably still connected to economic recession.

Such a conclusion may be drawn after consideration of the ratio of CO_2 emissions (from energy-related sectors) to GDP – the carbon intensity (see

---

[*] The following remarks should be treated as an individual input to the discussion on these important issues raised by COP 1 and reflected in the Berlin Mandate.

Figure 6.1: GDP changes in the EIT countries in the years 1990–93

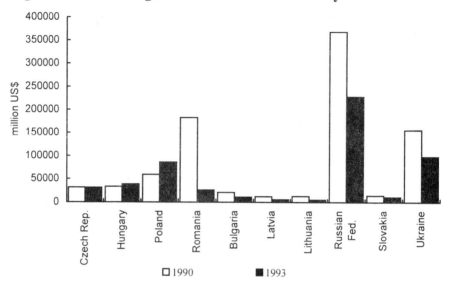

Source: Author.

Figure 6.2: Energy-related CO$_2$ emission changes in the EIT countries

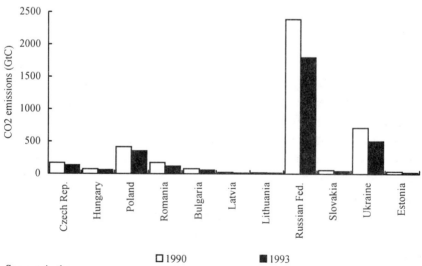

Source: Author.

Table 6.1: Energy-related CO₂ emission changes, 1990–93 (Gt CO₂)

	1990	1993	%
OECD	12114	12289	1.4
EIT	4136	3118	-24.5
Annex I Parties	16250	15407	-5.2

Source: Author.

Table 6.2). In the first group of countries the ratio shows a decreasing trend, though the level of carbon intensity of GDP is still approximately four times higher than in some OECD countries. The exception is Hungary, where this value is comparable with some EU countries. A level of carbon intensity similar to that of OECD countries would be reached by most EIT countries in the next century. Such a process occurring in Poland is shown in the scenario in Figure 6.3. It is expected that under a scenario involving measures directed towards reducing GHG emissions, such as a carbon tax, this level will be reached by Poland in the 2020s. In the second group of countries the carbon intensity is increasing since the decline in GDP is greater than the reduction in CO₂ emissions.

In the negotiation process another aspect of the situation of EIT countries should be taken into consideration. Some EIT countries have become members of the OECD (the Czech Republic and Hungary), and another (Poland) will be a member soon. Moreover, these countries, and maybe others too, are applying to be members of the European Union, and will probably join at the beginning of the next century. The question is for how long, after becoming a member of the OECD, and the EU in particular, may a country be considered 'an economy in transition' and allowed to use Article 4.6 of the Convention, which gives it flexibility in meeting its commitments on emissions reduction? If a newcomer declares itself to be a developed country, then its commitments should be similar to those of other similar Parties. However, expectations that such decisions will be declared before COP 3 are not justified. In such a case, one of the most reasonable options would be to establish two separate protocols for both groups of EIT countries mentioned above. Of course, both protocols should be open for each EIT Party. Another option is for some EIT Parties voluntarily to share commitments with OECD or EU Parties.

Table 6.2: Carbon intensity – ratio of CO_2 emissions to GDP

OECD	1990	1993	EIT	1990	1993
Australia	0.91	0.92	Czech Rep.	5.25	4.23
Austria	0.37	0.34	Hungary	2.18	1.58
Belgium	0.57	0.57	Poland	7.04	4.09
Canada	0.76	0.77	Romania	9.37	4.54
Denmark	0.41	0.44	Bulgaria	3.72	5.25
Finland	0.39	0.5	Latvia	1.92	2.01
France	0.32	0.31	Lithuania	1.41	3
Germany	0.65	0.56	Russian Fed.	6.49	7.89
Greece	1.1	1.07	Slovakia	4.05	3.8
Ireland	0.77	0.7	Ukraine	4.57	5.14
Italy	0.38	0.37			
Japan	0.36	0.35			
Luxembourg	1.22	1.25			
Netherlands	0.58	0.58			
New Zealand	0.57	0.63			
Norway	0.3	0.28			
Portugal	0.7	0.76			
Spain	0.44	0.45			
Sweden	0.23	0.24			
Switzerland	0.19	0.19			
Turkey	1.28	1.2			
UK	0.6	0.57			
USA	0.9	0.89			

Source: Author.

The quantification of targets and the criteria for emissions reductions are the most crucial issue, and not only for EIT countries. However, for these Parties the question is especially difficult, as the majority of them are not yet able to foresee the future development of their national economies. It creates problems concerning the estimation of future levels of GHG emissions, and undertaking any specific obligations might involve some risk of exceeding the adopted level of emissions.

Taking these factors into consideration, it seems that two approaches to the question may be proposed. First, on the basis of scenarios developed for their national GHG emissions limitation strategies, countries would declare the possibility of their acceptance a target of

Figure 6.3: CO₂ emission intensity of GDP in Poland: alternative paths of economic development

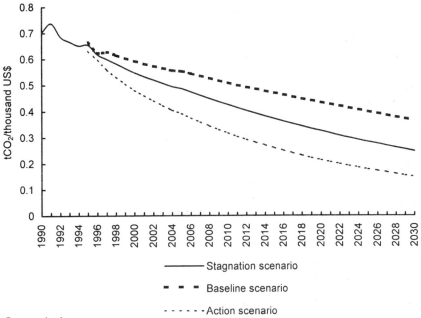

Source: Author.

reduction/stabilization of emissions under an established protocol for the group of EIT Parties. It seems that at least until 2005 it is possible to estimate the projected level of emissions reasonably accurately, at least for CO_2. Another approach, which makes sense only when applied to all Annex I Parties, is that the reduction level will depend on the proportion of a country's share of Annex I Parties' total emissions. Table 6.3 illustrates this idea.

Of course, instead of total emissions for Annex I Parties, emissions for the world as a whole could be taken as a basis for sharing commitments. It seems that after 2005 this may be necessary, as some developing country Parties may join the Annex I Parties in their efforts to reach the goal of the Convention.

Starting from emissions levels in 1990, EIT countries could plausibly reduce their emissions by approximately 11% by 2005. For the next target

Table 6.3: EIT share of 1990 total emissions of energy-related CO_2 of Annex I Parties

Country	Emission (%)	Reduction (kt)
Russian Federation	16.26	378,429
Ukraine	4.96	35,300
Poland	2.90	12,022
Romania	1.18	1,994
Czech Republic	1.10	1,731
Bulgaria	0.54	414
Hungary	0.48	324
Slovakia	0.38	211
Estonia	0.26	96
Latvia	0.16	36
Total	28.22	430,557

Source: Author.

year – 2010 – the base year for emissions-sharing should be 2000; for 2015, it should be 2005, and so on. Such a progressive approach will create natural incentives for emissions reduction, because a country will be interested in contributing as little as possible to total emissions.

Chapter 7

Approaches to differentiation for advanced developing countries

Hoesung Lee

The notion of 'common but differentiated responsibilities' is typical of those concepts to which agreement on principle is easy but agreement on action proves controversial. Considerations of equity play a pivotal role in these controversies.[1] Historical responsibility for causing climate change, though opposed by some countries, was one of the criteria used in delineating the Annex I group. The criteria by which developing countries participate in the global reduction of greenhouse gases remain to be determined.

The principles contained in the FCCC are all-encompassing, and thus invite more contention than clarification. The need for global action to counter climate change is set alongside the need for economic development. The appropriateness of global action should be assessed according to the local conditions of each party. Policies and measures are to be integrated with national development programmes, while 'economic development is essential for adopting measures to address climate change'. The FCCC allows each party to approach climate change problems from its own vantage point.

The energy and environmental policies in individual countries reflect the culture, institutions and politics prevailing in each country.[2] The history of the energy and environmental policies in Korea reveals the following characteristics:[3]

- The precautionary principle was never a factor in the development of energy/environmental policy. The policies were developed after the fact

---

[1] IPCC, *Equity and Social Considerations Related to Climate Change*, Proceedings of IPCC WGIII Workshop, Nairobi, 1994.

[2] Michael Grubb, 'Energy policies and the greenhouse effect: a study of national differences.' *Energy Policy*, 19:10, December 1991.

[3] IEA, *Energy Policies of the Republic of Korea*, International Energy Agency, Paris, 1994.

in response to extreme events such as the oil shocks and severe pollution in drinking water.

- The goal of energy policy was to ensure the flow of supply to support the industrialization programmes. This is how the petroleum refining industry began thirty years ago, and it has become a dominant energy supplier. The diversification of energy sources in later periods also reflects the supply-security-first policy.
- The external shocks triggered the introduction of alternative energy sources. The first oil shock brought about the nuclear electricity development programme. The second oil shock led to the launching of the liquefied natural gas (LNG) import programme.
- The government was the decision-maker in every phase – energy resource acquisition, pricing, investment and strategic planning.
- The rules and regulation approach was the policy of choice. Markets were distrusted both by the government and by the private energy supplier.
- In environmental policy, large gaps exist between regulations/standards on the one hand and implementation and performance on the other. Policies were designed as a result of pressures from environmental groups, usually in response to some measure of environmental crisis. Industries found little motivation to implement the stringent environmental policies, thus leading to gaps between the policy and reality.

Climate change policies will be developed within the sphere of these fundamental characteristics in the development of energy/environmental policy. Policy-makers will not respond to calls for actions based on the precautionary principle. It is only extreme events of catastrophic proportions that will cause them to take action. Otherwise, the specific climate action programmes will be so designed that reduction of greenhouse gases will not be in conflict with the national goal of achieving economic and social development as well as the national aspiration of improving energy security and environmental protection. Three possible courses of action exist to satisfy this requirement when additional commitment is considered for advanced developing countries such as Korea. These are OECD income parity, commitment to efficiency improvement, and energy subsidy removal.

OECD income parity

The rationale for this criterion is the principle of ability to pay. Newly industrialized countries should not be expected to become involved in emission constraints like those borne by the OECD until their per capita income reaches some parity with the developed countries. However, this criterion should be forward-looking: if an NIC's income is expected to rise to reach parity with current OECD income levels over the period for which emission constraints are being considered, it could be legitimate to start considering now appropriate constraints over that timescale, but not to a degree comparable with those of the already developed countries.

If, as the Climate Convention requires, the developed countries reduce CO_2 emissions by 2000 to the 1990 level, then the income level of those countries in 2000 could be a threshold for commitment by NICs. If the current commitment is to be fulfilled some years after 2000, the threshold income level will be accordingly higher. Korea expects its income to rise by 2010 to the current level of OECD average income, implying that the appropriate timing of commitment will be 2010 at the earliest.

This does not mean that no action will be taken in the intervening years. Rather, the expectation of future commitments for CO_2 reduction will induce businesses to factor in the new constraint and to find cost-effective strategies early in their planning and investment. To reduce business's uncertainty for business over the timing of commitments, the government may wish to promulgate far in advance a set of conditions which will lead to commitments being undertaken. Obviously, income parity will be the primary condition.

Commitment to energy efficiency improvement

What is known as the Kaya identity indicates four elements contributing to increases in CO_2 emissions: the carbon intensity in energy consumption, energy intensity, per capita income, and population.[4] The reduction in energy intensity reduces CO_2 emissions at little or negative cost. Opportunities

---

[4] Y. Kaya, 'Impact of Carbon Dioxide Emission Control on GNP Growth: Interpretation of Proposed Scenarios', IPCC Response Strategies Working Group, Geneva, May 1989.

for energy efficiency improvement are more readily available in developing countries than in developed countries. Accordingly, developing countries may contribute to the global reduction of CO_2 emissions by committing themselves to the specific target of efficiency improvement to the extent that such a commitment may be beneficial to economic development.

The commitment to efficiency improvement induces countries to pursue energy-efficient and less carbon-intensive economic development strategies. Such a commitment will be easily accommodated because it frees the country from an aggregate limit of CO_2 emissions. In fact, it allows the country an entitlement to development-related CO_2 emissions as long as the development is energy-efficient and thus less carbon-intensive. In terms of the Kaya identity, this commitment and the current commitments of developed countries complement one another to reduce overall CO_2 emissions in different ways.

Energy subsidy removal

Developing countries may adopt a programme of subsidy removal as a commitment to help a global effort to reduce CO_2 emissions. The pros and cons of energy subsidy removal are well known. The improvement of economic efficiency is the major advantage of subsidy removal. However, such an action may provoke a political backlash. Thus the challenge is to assemble and nurture the constituency for market reform.

Establishing international energy efficiency standards for certain manufactured products would facilitate building up such a constituency. Suppliers of energy-efficient products will support the removal of market distortions such as subsidies. So will the proponents of stronger measures for environmental protection. The international efficiency standards will strengthen the positions of these groups.

Consequently, there could be scope for NICs to be involved in some policies and measures that may be established under the Convention, some time before they could equitably be involved in emission targets. Indeed, if the policies and measures were seen to establish useful guidance for best international practice, this could make it more attractive for a number of the more advanced developing countries to become involved on this basis, and this might also assume relevance with respect to their application for OECD membership.

For developing countries, economic development and the improvement of the local environment are more urgent than problems of global climate change. Actions to mitigate climate change will make sense only if they facilitate economic growth. The three alternative approaches to climate actions commitment analysed here are measures that meet the growth requirement.

Chapter 8

Differentiation among non-Annex I Parties and the potential evolution of international commitments beyond the Berlin Mandate negotiations

Diego Malpede

The Berlin Mandate

During the final stages of negotiation of the Conference of the Parties, minutes before two groups of negotiators crafted the final terms of the Berlin Mandate, several environment ministers from developed countries expressed to their developing country counterparts their need to return home with a signal or a message for their citizens that all countries would be making the necessary sacrifices to address climate change.

With that in mind, the negotiators undertook the last effort to reach a compromise. The use of the word 'process' instead of negotiation allowed the OPEC representative to accept the wide recognition that the commitments under Article 4.2(a) and 4.2(b) were not adequate to achieve the overall objective of the Convention.

Developing countries then put pressure on the use of the word 'targets', following the lead taken during the Conference by the so-called 'Green Group', that is to say, the G-77 and China, minus OPEC. The outcome of the deliberation within Annex I countries on this question was long awaited, revealing the different positions that existed.

At that point, after the Annex I Group refused to accept the word 'targets', the expression 'quantified emission limitation and reduction objectives' appeared. Many people outside the Conference, especially environmental NGOs, would have preferred 'targets', but although qualified by the word 'objectives', the concept of 'reduction' is nevertheless present. It seems that the JUSCANZ countries[1] were absolutely firm in the setting of several time horizons instead of only one.

---

1 Japan, the USA, Canada and New Zealand – a loose forum for coordinating positions in climate negotiations.

During the last bargaining hours of the morning, negotiation focused on the request made to developing countries to share in the global effort. The major developing countries rejected once again the possibility of accepting any kind of new commitments; they even the expression 'enhance the existing commitments'. So the final wording states that all Parties (i.e. including developing countries) 'reaffirm existing commitments in Article 4.1 and continue to advance the implementation of these commitments in order to achieve sustainable development, taking into account Articles 4.3, 4.5 and 4.7'.

This is a brief description of how the most important issue on the agenda of the First Conference of the Parties, the review of the adequacy of commitments, was negotiated, and gives a flavour of how difficult it would be to differentiate the commitments of various countries in the implementation of the Convention. The reference in the final wording of the Berlin Mandate to 'sustainable development' is a very important one, because it can provide a possible basis for consensus, so difficult to achieve given the well-known North–South tensions underlying the Climate Convention.

The North–South cleavage

When these tensions surface – and they can appear at any moment of any meeting regarding any of the issues on the agendas of the subsidiary bodies of the Convention – the attitudes of political confrontation based on sovereignty principles tend to push aside more cooperative positions.

Let me give a few examples. First, the recent agreement on a reporting framework for Joint Implementation, a highly controversial idea, only occurred after agreement that its name be changed. Because many important developing countries did not like any implications related to shifting of responsibilities regarding mitigation commitments, Joint Implementation is now called 'activities implemented jointly'. The final negotiation on the reporting system for this type of project was close to failure because of the insistence of the OECD countries on the incorporation of the 'monitoring' requirement, an essential tool to verify the reduction of emissions in a project. The word sounded too intrusive for some delegations. The wording finally agreed upon is 'mutually agreed assessment procedures'.

More important still has been the issue of communications by non-Annex I countries. The Conference of the Parties or its subsidiary bodies must decide on guidelines for a format for reporting national communications. At the outset it is essential to recognize that most developing countries do not have either the economic or the human resources to fulfil this very complex obligation comprehensively, nor fora like the OECD, the International Energy Agency or the Annex I Experts Group to coordinate and exchange information between them. The accomplishment of this important task lies outside the ambit of the normal functions of the G-77.

It seems that developed countries want to see the establishment of an extensive, comprehensive reporting system on greenhouse gas emissions, similar to what they themselves have undertaken since the Convention entered into force. The aim would be to ensure that everybody knows what other Parties are doing, even those with different obligations, in order to permit the competent bodies to assess the overall aggregated effect of steps taken to implement the Convention. Some influential developing countries, however, do not want a deep scrutiny of their environment and development policies, preferring instead a very simple communication, with little scope for disaggregation of data and detailed descriptions of actions which address climate change. Another difficult issue is the existing lack of agreement on the criteria for deciding the composition of Technical Panels on Technology and Methodologies, two fundamental issues for all countries.

Although many of these controversies arise from reasonable and legitimate concerns, some are not well founded. In general, all of this is exhausting for the negotiators and is usually a waste of time and energy.

Emission realities and burden-sharing

One of the basic principles of the Climate Change Convention, and also of the Rio Declaration and other legal instruments, is 'common but differentiated responsibilities'. This phrase is also quoted in the preamble of the Berlin Mandate. How many common responsibilities and how many differentiated ones will be agreed in the final outcome? Between whom? It is very clear from the language in the Mandate that there will be no mitigation commitments for developing countries.

The question of who created the problem in the first place is also clear. The huge difference in per capita emissions between developed and developing countries, which will continue for many decades, is too large to be ignored. In addition, these differences in per capita emissions also exist within the borders of the majority of countries, where the conspicuously high consumption of many people contrasts with the realities of poverty and social exclusion for many others. This substantial disparity in per capita CO_2 emissions will continue into the next century.

All the reports reach similar conclusions about the existing and potential increase of atmospheric concentrations of GHGs, in particular CO_2. The International Energy Agency's 1996 *World Energy Outlook* describes the level of emissions and projections in terms of two scenarios: one where there are significant constraints on capacity (generating or supply), and one where there are significant attempts to save energy.

The world's rate of energy growth, which reaches 19% per annum in some countries, is fuelled by population growth, economic development on an energy-intensive industrial and agricultural basis, and urbanization. Power-generating capacity alone doubled in the 1960s, again in the 1970s and again in the 1980s.[2] Rates of growth in energy, however, vary greatly across countries, with OECD countries often growing at less than 1%, and the most rapid growth coming from developing countries, particularly newly industralized countries. By 2010, in terms both of capacity constraints and of energy saving, total CO_2 emissions from non-OECD countries are likely to have overtaken those from the OECD, and China will remain the largest single source of emissions and is projected to more than double its emissions by that same date.[3] This could happen before or after that date, but it will happen. So it is very clear that the efforts developed countries should make will not be enough. This could be seen by some as ignoring the emissions per person and historical considerations, but keeping differentiation as it currently is in the first two or three decades of the next century will not solve the problem in the long run.

---

[2] V. Bakthavatsalam, 'Financing India's Renewable Energy Boom', presentation at the Conference 'Changing Politics of International Energy Investment', convened by the Royal Institute of International Affairs, London, 4–5 December 1995. Dr Bakthavatsalam is Managing Director of the Indian Renewable Development Agency.

[3] *World Energy Outlook*, IEA, Paris, 1996, p. 60.

One of the main concerns of developing countries, which have the need and right to grow, to expand their industries and to secure the human development of larger and more urbanized populations, is the fact that the Annex I countries have apparently failed to take the appropriate measures to ensure a return to the level of 1990 emissions by the year 2000. This perception, as well as other uncertainties arising from the first national communications under the Convention, such as the 'adjustments' artifice used by some developed countries in their first national communications under the Convention, gives a credible moral standing to the position of the G-77 and China – quite apart from the general perception in developing countries about how slowly monetary resources are flowing to the financial mechanism of the Convention.

The 'double speech' syndrome

This North–South division clearly disrupts what should be a global and constructive cooperative effort to address a common good. What is remarkable and curious is the observable 'double speech' in the climate regime; that is to say, the contradiction between the controversial political perceptions and negotiation positions adopted by developing countries, and the great amount of national and international cooperative activities aimed at mitigating climate change in developing countries. The number of studies and research activities designed to estimate current GHG emissions as well as technological and policy options for mitigation, including the economic, social and environmental cost of implementing such options, is increasing daily in developing countries.

Several national and regional studies have been initiated during the past three years by international and regional organizations such as UNEP, the Global Environment Facility (GEF), the UNDP, the World Bank, the Asian Development Bank and the Inter-American Development Bank. An enormous amount of work is also being done at the bilateral level, such as the programmes of the United States, the European Union, Japan, Australia, Canada and others, as well as by many private institutions, the business sector and associations and NGOs.

The list of energy-related projects in developing countries and economies in transition under consideration for financing by three large multilateral development banks (the World Bank, the Asian Development Bank and the Inter-American Development Bank) can be obtained from the Internet site of International Institute for Energy Conservation. This database shows 315 projects. The vast majority of them, once implemented, should result – directly or indirectly – in a decrease of GHG emissions.

The US Environmental Protection Agency has signed a protocol of cooperation with China, comprising many environmental actions in fields such as health, pollution control, environmental processes and effects, and environmental management (which includes a GEF project on coalbed methane recovery and utilization). These two governments also agreed on a series of climate-change activities, including the establishment and functioning of the Beijing Energy Efficiency Center, support for the Chinese Country Study, Cooperative Research on Climate Change, Liaoning Province Greenhouse Gas Inventory and Options for Reduction, and the projects focused in China under the US Technologies for International Environmental Solutions (USTIES), which include renewable energy projects, gasification and pollution prevention.

The burning of immense quantities of coal, oil, gas and fuel wood also constitutes a serious problem of local pollution. The sulphur and nitrogen oxides which result from the burning of fossil fuels are the main cause of smog, acid rain and a variety of serious effects on the environment and human health. China's Ministry of Coal has appealed for foreign research funds to develop low-emission 'clean-coal' technology, given the difficulties of developing alternative energy sources.[4] The Chinese government, in the report 'Issues and Options in China's Greenhouse Gas Emission Control', funded by the GEF and carried out by the World Bank, UNDP, and the National Environmental Protection Agency and State Planning Commission of China, admits that the country could suffer permanent floods with huge population displacement if major steps are not taken to limit carbon dioxide emissions. This report outlines a two-pronged strategy to help China reduce GHG emissions: on the one hand, a set of priority

---

[4] *International Environment Reporter*, 1 May 1996, p. 380.

investment and technical assistance programmes to encourage the adoption of more efficient and low-carbon energy technologies and to improve institutional and human resources in industry, partially funded through international development organizations and future GEF resources; and on the other, the continuation of the economic reform process whereby market incentives improve resource allocation.[5] Finally, in 1995, the United States and India signed the 'Common Agenda' project, which plans to help India to curb greenhouse gas emissions from electricity generation, concentrating it in biomass fuels for cogeneration in the sugar industry and on improved coal combustion techniques for the thermal power industry, among many other actions.[6]

If these important developing country players are trying to formulate and implement policies aimed at reducing dependence on fossil fuels, why is it so difficult to make progress at the diplomatic level? Effective cooperation to address climate change would need to overcome the obstacles stemming from sovereignty concerns. It is one thing to say: 'I'm trying to implement sustainable development but I have these concrete problems, and we would need financial and technological assistance for these sectors and regions...', and afterwards try to construct rational ways of action. But it is quite another matter and less satisfactory to say: 'I don't want to talk about or refer to these issues, because it is a domestic affair.' Unfortunately, the latter is commonplace in negotiating fora.

Towards differentiation?
The differences among developing countries in terms of stages of development, endowment of natural resources, fossil fuel consumption, competitive advantages and socio-political cultures are far greater than those which exist between the Annex I countries. This complicates any discussion of differentiation between them, but also suggests some possible lines along which differentiation might be organized.

---

[5] *International Environment Reporter*, 19 April 1995, p. 300.
[6] *International Environment Reporter*, 3 May 1995, p. 346.

A first differentiation could be made easily by separating from potential mitigation involvement the least developed countries and heavily indebted low-income countries, even though these countries can also implement successful environmental actions.

A number of countries are approaching fulfilment of the political and economic criteria for OECD membership. They could conclude (or be persuaded to conclude) that it is desirable, both for the global and local environment and the economy itself, to try to deepen their sustainable development policies and take advantage of, what the IPCC calls 'significant no-regret opportunities', applying cost-effective technologies and policies to markedly reduce their net emissions of GHG from industrial energy supply, energy demand and land management practices.

The categorization could be based on many factors, including GDP, total emissions, GDP and emissions per person, human development index, carbon efficiency, or others. Possible lessons concerning criteria to be used for Annex I differentiation could also be applied here. It would be useful to speculate on the nature of the involvement of such countries in commitments under the Convention. Thus, the decision to mitigate could adopt the following forms or scenarios.

Voluntary involvement

This is a possibility already contemplated in Article 4.2.g of the Convention. As a hypothesis, this formula could be redrafted and updated for the final text of the Berlin Mandate Protocol or for a future solution. The relevant part of the Berlin Mandate, which would fit perfectly here as a difficult framework for negotiations, is the phrase 'continue to advance the implementation of existing commitments'. Here there is room to provide some type of incentives to undertake commitments, such as additional assistance or cooperation rewarding willingness to join the regime. This might be unacceptable for the countries not wanting to be part of it, because it would act as a tacit negative conditionality. A non-Annex I country might also decide to undertake mitigation policies without making this declaration, and go along outside the institutional machinery of the Convention, with or without new incentives. Finally, if there are Annexes to the Protocol on Sectorial Policies and Measures, developing countries with specific economic advantages in any particular sector could adhere to particular annexes to which there was a strong likelihood of their compliance.

Legal obligation

The drafting of new types of obligations for a particular category of countries would be the most difficult option, because of the North–South tensions mentioned above. According to the precedent provided by the Convention, an Annex to the Protocol would be the way to frame a decision about which countries would join such an obligation. The final formulation of the list of members of this new Annex will obviously need their agreement to be effective. But the Berlin Mandate language does not provide a strong legal basis for the elaboration of new commitments for non-Annex I countries before the meeting in Japan in 1997.

Other alternatives

In 1995, in this same forum, Michael Grubb mentioned the possibility of linking mitigation commitments to accession to the OECD.[7] This could be a promising idea (leaving aside the potential dislike of the concept of conditionality this could provoke among many countries), but the recent practice of the Convention raises many doubts. In fact, we see the opposite, with a real risk of OECD members leaving the ship. One such case could be Turkey, which is still not a Party to the Convention, and which has remained silent about its intentions. Another possible case is Mexico, a country with very active participation in the Convention until it joined the OECD. The Mexican case may become a pilot for this idea. There are no signs of possible discussion within the OECD or between Annex I countries about these situations.

I understand that current and future commitments under the Climate Change Convention are a very important restraining factor, affecting, among others, South Korea's decisions on whether to join the OECD. All these issues should be analysed in the general framework of standards of accession to the OECD.

---

[7] Michael Grubb, 'From Rio to Kyoto via Berlin: Climate Change and the Prospects for International Action', in Michael Grubb and Dean Anderson (eds), *The Emerging International Regime for Climate Change: Structures and Options after Berlin*, Royal Institute of International Affairs, London, 1995.

The possibility of linking mitigation commitments to accession to the OECD is also very difficult to analyse in the context of the ongoing debate about the future of the OECD itself. The former Secretary-General of this organization said last year that 'if an increasing number of non-member countries become possible candidates, we will have to face the problem of the long-term orientation of the Organization. It would be necessary to be prepared, one day or another, for the possible accession of Russia, China, India or Brazil.'[8]

Avoiding approaches which raise problems concerning conditionality, James Sebenius, from the Harvard Business School,[9] proposes that a small group of developed countries, probably the EU and JUSCANZ, negotiate emissions reduction targets and timetables among themselves. By avoiding the North–South clash, this agreement would be likely to prove far easier to achieve than a global accord, and, if the provisions were serious, it would give high moral standing to its members. This smaller-scale agreement could be expanded later with the use of incentives to bring in key developing countries such as China, India, Brazil, Indonesia (Sebenius also mentioned Mexico), as well as EIT countries. The incentives could be resources obtained from a carbon tax among the Annex I members of the group or a stronger replenishment of international development assistance. Negotiations with a particular newcomer country could set, for example, a schedule of emission targets and provision of significant incentives to reach them, such as assistance in developing and exploiting natural gas reserves, improvement of coal use or transfer of the latest technologies.

Sequence

These ideas on possible involvement, with all the inherent uncertainties, must be analysed within the existing time-schedule provided by the deadline of having to negotiate the Berlin Mandate by COP 3. One possible out-

---

[8] Speech by Jean-Claude Paye, Secretary-General of the OECD, at the Parliamentary Assembly of the Council of Europe, Strasbourg, 28 September 1995.
[9] James K. Sebenius, 'Towards a Winning Climate Coalition', in Irving M. Mintzer and J.A. Leonard (eds), *Negotiating Climate Change: The Inside Story of the Rio Convention*, Cambridge University Press, Cambridge, 1994, pp. 277–320.

come could be a combination of some of the variants. But I do not believe there is enough time to develop all these ideas sufficiently to find a constructive trade-off for all Parties involved before the deadline of late 1997. Developing countries with strong positions will make every effort to reject any linkage between a stringent Annex I emissions reduction protocol and subsequent involvement from developing countries.

Even with good ideas and a lowering of the level of confrontation, COP 3 is very close. For example, take Argentina. My country shows an efficient environmental record in terms of energy supply and decrease of emissions in recent years. In our delegation we always discuss potential involvement in mitigation policies, seeing the Convention as an 'opportunity' for clean growth with 'no regrets' policies, convincing the private sector that industrial operations could be clean, profitable and attractive to foreign investment at the same time. But the fact is that we should wait for the completion of the inventory to see the picture more clearly, in order to know where we stand. The inventory and the national communication of Argentina must be presented in June 1997 (with a similar deadline for other developing countries), and clearly there does not leave enough time to decide and fully consult among the many agencies involved, the Congress and the private sector about a political decision of that importance before COP 3.

Thus, it would be prudent for us to wait until we see changes in the positions of developing countries – which, to a great extent, depend on what developed countries will do. At some point, after the Kyoto meeting, there could be a better negotiating environment to allow a focus on necessary and creative ways to find a different role for developing countries.

In the meantime, many suggestions, if implemented, could contribute to a better atmosphere in which countries avoid 'playing politics' or forming blocking coalitions, and try to cooperate in the strengthening of the Climate Convention. Developed countries must demonstrate once and for all their so often quoted 'leadership', with clear actions that could give them some moral authority to ask for mitigation on the part of other countries. These could include:

- Real progress in the Berlin Mandate negotiations (for example, narrowing down effectively the number of policies and measures and refraining from delaying tactics).
- Lobbying to persuade OPEC countries to solve once and for all the regrettable failure to adopt the rules of procedure.
- Establishing a cooperative framework of action with the issue of national communications of non-Annex I countries.
- Facilitating and promoting technology transfer mechanisms to developing countries aimed at energy conservation, improvement in energy efficiency and the shift to low carbon fuels; encouraging private sector and financial multilateral institutions to invest in these sectors and technologies.
- Showing clear signs of willingness to change the structural status quo, e.g. showing determination to establish international carbon taxes with possible potential transfers of financial resources, funds and technology to developing countries. Any other sign that could contribute to changing damaging consumption patterns in the medium or long term, such as subsidies to fossil fuels or the transportation sector, will contribute to changing legitimate perceptions in the South.

At the same time, developing countries must also change attitudes and perceptions. One very useful action would be to avoid the 'double speech' syndrome mentioned earlier and keep the negotiation position in tune with national actions and efforts – for example, those directed at the creation of sound economic and legal conditions for investment in clean technologies. At the same time, in the negotiations, they must make every effort to avoid confrontation, knowing that we are all passengers in the same boat. Mitigation efforts would also be beneficial for social and economic development. Sustainable development should act as the guiding motivation on all fronts.

Finding solutions to the attainment of the objective of the Convention must necessarily involve continuous global efforts, based upon enlightened self-interest. Ethics must have a practical meaning and sense, aimed at ensuring the long-term survival of humankind.